Hey, UX!

10 things your **Product Manager** wants you to know.

A book about **building great products** through **great partnership.**

JEFF CARPENTER

Special Thanks to:

Brad Barnhart
Amy Battles
Cindy Brummer
Katie Cole
Kamala Espig

10 things your Product Manager wants you to know:

#1   Be Curious About Me!

#2   Understand How I Think

#3   Recognize I am Stretched Thin

#4   Appreciate I Have Limitations

#5   Be Proactive; Offer Solutions

#6   Speak My Language

#7   Get to Know the Business

#8   Let's Show Business Value Together

#9   Provide On-Time Deliveries ...with Alternatives!

#10   Offer Research and Reasoning

Bonus: A PM-UX Partnership Manifesto

Introduction

Great Product Managers and UX Designers have one thing in common: they are both excellent problem solvers. They don't get distracted by things that don't matter. They stay focused on customer needs and how to solve them!

Yet, even with this shared mindset, Product Managers and UX Designers often struggle to work seamlessly together. Between them, there can be tension, pressure, conflict, disappointment, and unmet expectations. UX Designers often feel that PMs prioritize speed and business objectives over user experience, leading to rushed timelines and untested designs. They can feel frustrated when their input seems undervalued or when they're brought in too late to make meaningful contributions. Meanwhile, PMs sometimes see UX Designers as overly focused on ideal solutions that look great, but don't fully account for technical constraints, deadlines, or business priorities.

When great Product Managers and UX Designers collaborate effectively, they help everyone around them stay focused by continually asking the essential question: *"What problem are we trying to solve?"* So, when it comes to this book, let's take the same approach. Let's ask: What is the problem that keeps PMs and UX Designers from being great partners? Why do so many teams experience conflict, disappointment, or mismatched expectations? Is it a lack of process, skill, training, unclear roles, poor communication: or something else?

Unfortunately, there isn't a single, overnight fix or silver bullet for creating a great UX/PM partnership. A better intake process, handoff mechanism, or meeting structure won't magically create a perfect partnership. But what you, as a UX Designer already innately possess as can make a big difference: **empathy**.

Empathy is at the core of what makes you a great UX Designer. It's why you chose this field: because understanding others' perspectives, pain points, and struggles comes naturally to you. You thrive on finding creative ways to ease those pains, improve lives, and bring joy to others. Great products aren't created in isolation. They require multiple people and perspectives coming together to succeed. The same empathy you have for your users can be directed toward your team, and specifically, your PM partner.

Building a strong partnership isn't about putting the responsibility solely on you, nor is it about perfection. It's about taking small steps to better understand one another and support each other. By channeling your empathy toward your PM: understanding their goals, constraints, and pressures: you can build trust that leads to better collaboration and great outcomes for everyone: your users, your business, and your team.

> The more you invest in this partnership, the more you'll find yourself with the time, resources, and respect you need to do what you do best: create memorable experiences that improve users' lives.

Your partnership won't be perfect at first: it will evolve over time. And over time, focusing on problem-solving, mutual respect, and trust, you'll not only build a better working relationship, but also build great products that delight customers and propel both your careers forward.

This book is one half of a larger conversation. For a PM and UX Designer to truly thrive together, both must embrace empathy, curiosity, and collaboration. This isn't about pointing fingers or delegating blame: it's about understanding and supporting each other to build great products together.

With that in mind, here are ten things Product Managers want you to know, *in their words*.

#1
Be Curious About Me!

Be interested in me and my problems,
as much as you are Users.

As a Product Manager, I understand and appreciate the vital role UX Designers play in ensuring our products meet user needs and expectations. Your dedication to understanding users' problems, building personas, shadowing them, and empathizing with their needs is invaluable. To create truly successful products, we need to balance user-centric design with the broader business objectives and challenges that I face daily.

To support our shared goals, get curious about the challenges I navigate as a PM. What keeps me up at night? What are the key performance indicators (KPIs) that define our success? Are there market trends, customer feedback, or competitive pressures influencing my decisions? How do these factors shape the product roadmap and timelines? When you understand these dynamics, we can work better together *within* those constraints to create designs that delight users and align with business goals.

Consider creating a Persona for me, your PM! What drives me? What frustrates me? What does success look like for me and my career? What are my top pain points? What bottlenecks do I face? I might be juggling tight deadlines, limited resources, or the challenge of balancing short-term goals with a long-term vision. By understanding these pressures, you can offer design solutions that are not only innovative, but also feasible and strategic.

Remember, I rely on your expertise to make the best design choices. But those choices aren't decided in a vacuum. A feature that improves user experience but significantly increases development cost could disrupt timelines, budgets, or overall product viability. By engaging in open conversations about trade-offs, we can find middle ground where user needs are met and business goals are supported.

This isn't about shifting responsibility to you or asking you to compromise your design values. It's about collaboration. By being curious about my goals, constraints, and decision-making process, you'll gain a clearer picture of how your work impacts the product as a whole: and how we can succeed together.

In summary, be as curious about me, your PM, as much as you are about our end users. By understanding my challenges and goals, we can create products that not only satisfy users but also drive business outcomes.

Here are some things you can do to foster a deeper understanding of a Product Manager's role and challenges, leading to more effective communication and collaboration:

1. Shadow Your Product Manager

- Whenever possible, observe their daily routines and gain insight into their world.
- Watch how they make decisions, prioritize tasks, and communicate with stakeholders to align with overall business goals.
- Familiarize yourself with the tools they use, such as JIRA, Confluence, or Asana, to track progress, document decisions, and manage product strategies.
- Attend product management and cross-functional team meetings regularly. These meetings provide context around business priorities, constraints, and strategic goals.
- Make this an ongoing practice: not a one-time activity. As projects evolve, shadowing your PM periodically ensures continuous alignment and a deeper understanding of their evolving challenges.

2. Create Product Manager Personas

- Just as you create User Personas, develop a detailed Persona for your Product Manager.
- Just like User Personas, this PM Persona can keep you reminded of their perspective and constraints as you're designing solutions.
- Include elements like goals, pain points, challenges, and key relationships, to build a well-rounded understanding.
- Use these Personas to inform your interactions and design decisions, ensuring alignment with their priorities and constraints.

A Product Manager Persona might include:

- **Name**: Alex, Product Manager.
- **Goals**: Deliver feature X by Q4 to capitalize on market trends; reduce churn by improving onboarding flow.
- **Challenges**: Limited development resources; balancing quick wins with long-term strategy.
- **Pain Points**: Pressure from leadership for faster delivery; unclear product success metrics.
- **Motivations**: Creating a high-impact product that drives revenue while keeping the team aligned.
- **Key Partners/Influencers**: Director of Product Management, Engineering Lead, Director of Marketing.
- **Education:** Business Degree, currently acquiring a PM Certificate.
- **Career Track**: Previous Project Manager & Domain Expert.
- **Career Aspirations**: Become a Sr. PM within the year.
- **Working Environment:** Overall, above average PM Maturity level.

3. Co-Design Workshops

- Invite your PM to actively participate in the design process. Collaborative workshops are an excellent opportunity to brainstorm solutions together.
- Include other cross-functional team members, such as Engineering, Marketing, or Sales, who may offer valuable insights or ideas.
- Host these workshops during the *divergent thinking* phase of your design process, where generating a wide range of ideas is encouraged. This phase makes it easy to foster inclusivity and creative input from all participants.
- Later, during the *convergent thinking* phase, limit input and opinions from multiple parties. Instead use research and data to narrow down and validate design choices. This ensures decisions are based on evidence, not subjective opinions.

By applying these practices, you'll gain insight into your Product Manager's challenges, strengthen collaboration, and align designs with user and business goals, enabling informed decisions and impactful products.

2

Understand How I Think

I think in straight lines, not loops.
I don't get "Design is never done."

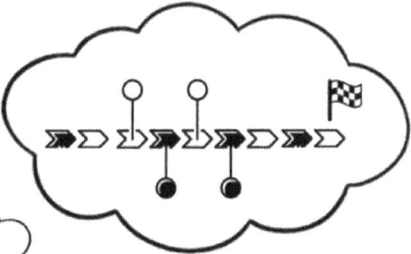

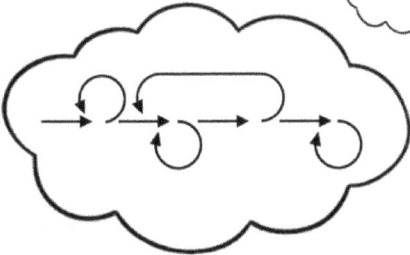

As a Product Manager, I understand and deeply value the unique perspective that UX Designers bring to the table. Your commitment to creating user-centered designs and your iterative approach to refining products until they are intuitive and delightful, are crucial to our success. However, the nature of our roles introduces different pressures and mindsets that can sometimes clash.

Product Managers often work under tight deadlines and are responsible for delivering projects on time, within budget, and aligned with stakeholder expectations. As PMs, our focus is hitting milestones and achieving specific goals, which requires a linear approach. We break work into phases, aiming to complete one step before moving on to the next. Often, we manage multiple parallel tracks of work, prioritizing and shifting our focus quickly, to meet business needs. This approach helps us manage risks, allocate resources, and keep stakeholders informed. Once a feature or version is released, our attention shifts to the next challenge, driven by the need to continuously deliver value and stay competitive.

I understand the iterative nature of UX design can sometimes feel at odds with the structured, linear approach of product development. Designers often see their work as an ever-evolving process, driven by user feedback and refinement, while Product Managers focus on meeting deadlines and delivering value within defined phases. These differing approaches can create friction, especially under time constraints, or when quick delivery is required.

To bridge this gap, we can align your iterative process with our linear, phased development model. One way to do this is through phased iteration: breaking the design process into smaller, manageable steps that fit within each phase of the release schedule. Instead of aiming for a perfect final design upfront, all in one fell swoop, we can deliver an initial, functional version quickly, then test it with users to identify major issues. We can then refine the designs based on feedback in subsequent phases. This approach allows us to continuously improve the user experience without derailing timelines. Over time, these incremental improvements build toward an optimal experience, while ensuring we stay on track.

Open communication is also essential. Regular check-ins and transparent discussions help us balance priorities, incorporate user feedback effectively, and navigate constraints together. I'll work to understand the rationale behind your design decisions and find ways to integrate them into the timeline, and I encourage you to collaborate with me to prioritize the changes that will have the most impact within each phase.

By aligning our processes and maintaining open dialogue, we can combine the strengths of iterative design with the focus and speed of phased development, and achieve the best of both worlds. We can deliver exceptional user experiences while maintaining the speed and focus required in our dynamic and fast-paced environment. This collaboration will not only help us release great products but also strengthen our partnership.

Here are some specific steps you can take to better align with a phased, linear development process and improve collaboration with Product Managers:

1. Establish Clear, Phased Milestones for Design

- Establish a long-term vision of what the user experience could be, to ensure the team is aligned and working in the same direction.
- Break the design process into manageable milestones aligned with the project timeline and implement changes in phases to enable iterative improvements without disrupting development.
- Define specific, achievable goals and deliverables for each milestone that balance immediate needs with long-term goals.
- Document how incremental changes build toward the broader vision, keeping the long-term strategy visible.
- Routinely communicate these milestones with the PM and development team to ensure alignment.

2. Conduct User Testing in Phases

- Schedule user testing at each project phase, rather than all at once, or waiting until the end.
- Use feedback from each round of testing to make incremental improvements that align with the current development stage.
- Share insights with the PM and team to prioritize actionable changes.
- Be open to shifting your design milestones based on user feedback or priorities

By aligning design iterations with phased development, fostering open communication, and focusing on shared goals, you'll help create a partnership that produces exceptional results for users, the business, and the team.

3

Recognize I am Stretched Thin

Understand that I am pulled in
different directions by many people.

As a Product Manager, my role involves balancing a multitude of competing demands from various stakeholders, each with their own priorities and expectations. I often find myself being pulled in multiple directions simultaneously, trying to address customer needs, meet business goals, and navigate the constraints of our technical teams. This constant juggling act can be overwhelming, which sometimes results in incomplete or imperfect communication, which I understand can be frustrating for you as a UX Designer.

For example, on any given day, I might start with a meeting with the sales team, who are eager to see new features that can help close deals with prospective clients. Immediately after, discuss technical debt with our engineering team and the need to refactor parts of our codebase to maintain long-term scalability. In between these meetings, I respond to urgent emails from the marketing team about upcoming product launches while also trying to gather user feedback from recent surveys to inform our next sprint planning session. Each of these interactions demands my full attention and input. Often, I have to make quick decisions without having all the information I'd ideally like to have.

To make things even more complex, the demands of one group often conflict with those of another, and my job is to balance these competing priorities while keeping the overall product vision intact. This is where your role as a UX Designer can complement mine. By understanding my challenges, you can help me create solutions that not only meet our users needs but also help balance the business and technical realities we are facing.

The risks I manage are significant. Prioritizing one feature over another can lead to dissatisfied customers or missed market opportunities. Delaying technical improvements might result in long-term maintenance issues, lost profitability for the business, or performance bottlenecks. Balancing short-term wins with long-term strategic goals is a constant challenge.

Focusing too much on new features to satisfy immediate sales targets might result in a product that becomes cumbersome to maintain and evolve. On the other hand, dedicating too much time to technical enhancements might slow down our feature delivery, impacting our competitiveness in the market.

Understanding these challenges can help you see why I might not always provide perfectly detailed design tasks or user stories for you to pick up and run with. It's a big part of why I often cannot give you the context or clarity to do your best work.

To help me out, and the entire team, I encourage you to be proactive, lean in, and engage with me throughout the product development process. Join me in stakeholder meetings, help gather requirements, and co-write user stories or acceptance criteria. Your active involvement can help bridge gaps and ensure we stay aligned in creating products that delight users while also meeting business objectives.

Here's how you can help ease the load and strengthen our partnership, ensuring users, their needs, and the overall user experience stay a priority despite competing demands.

1. Help Bridge Teams to Reduce Friction

- Act as a connector between sales, marketing, and engineering to ensure cross-functional alignment.
- Be an advocate: Help them understand and align on user needs.
- Become the go-to person for all questions about our users. Be the person people seek out when the wonder, "*How will our users feel about _____ ?*"
- Facilitate collaboration by considering the needs of each team while staying grounded in user-centered thinking.
- Share design updates early and seek feedback to address concerns before they become roadblocks.

2. Proactively Conduct and Share Research

- Offer to conduct usability tests, surveys, or competitive analyses to provide data when decisions are being made without enough information.
- Surface insights from user feedback, analytics, or support tickets to highlight problem areas.
- Use research findings to clarify conflicting priorities among stakeholders, helping us make more informed decisions.

3. Help Your PM Manage Stakeholder Expectations

- Attend stakeholder meetings to capture requirements directly and reduce the burden on me to translate between groups, or to you.
- Prepare talking points or presentations that help us both advocate for user-centered solutions during cross-functional discussions.
- Use data to justify design decisions, push back on unrealistic demands, and help us align expectations with reality.

4

Appreciate I Have Limitations

Some things are out of my control.
I'm not the CEO of my own product.

25

As a Product Manager, it might seem like I have ultimate control over our product roadmap, feature prioritization, and resource allocation. However, the reality is far more complicated. Many decisions are shaped, influenced, or even dictated by stakeholders such as leadership, investors, sales, and other departments. This external pressure often requires me to make trade-offs that might not be ideal from a UX perspective.

For example, there are times when leadership or investors push for the rapid development and release of a new feature to seize a market opportunity or respond to competitive pressures. In these cases, I may need to prioritize this new feature over improving existing ones, even if it means compromising on the thoroughness of user research or design polish. While I deeply value these aspects, the broader business context sometimes necessitates difficult decisions.

Additionally, budget and resource constraints frequently limit what we can accomplish. Even when we identify features that would greatly enhance the user experience, we might not have the resources or technical skills to make them a reality. For instance, while integrating a cutting-edge technology might be desirable, the cost or time required could conflict with our financial situation or project timeline. A big part of my role is balancing aspirations with feasibility.

Complicating matters further, I often navigate conflicting stakeholder priorities. Marketing might push for features to enhance campaign appeal, sales might demand functionalities that address existing client needs, and engineering might focus on reducing technical debt or improving system stability. My role is to balance these competing demands and their associated risks while maintaining the product vision.

If you've ever wondered, "*Why isn't my PM standing up for something they know is important?*" the answer might lie in these constraints. We don't have dominion over every decision or every stakeholder. As Product Managers, we must lead through influence, navigate organizational politics, and carefully choose our battles.

Sometimes, saying "*No*" too often erodes our credibility with leaders. Our job is to find ways to say "*Yes*". Pushing back too hard or too often eats away at our political capital and risks eroding trust with stakeholders, who expect us to balance competing priorities effectively in order to make forward progress.

Understanding these challenges helps you see why I'm not always able to immediately act on every design suggestion or piece of user feedback. Instead of viewing these limitations as roadblocks, let's work together to find creative solutions within the constraints. By collaborating closely, we can prioritize critical user needs that align with business goals, achieving the best possible outcomes given the circumstances. This approach not only strengthens our partnership but also ensures we're building a product that balances user satisfaction with business viability.

Here's how UX Designers can help navigate constraints and build a stronger partnership, help ease the pressure on your Product Manager, foster a more collaborative working relationship, and ensure that user experience remains a priority even in challenging situations:

1. Be Proactive in Identifying and Understanding Constraints

- Set up regular one-on-one meetings with Product Managers to discuss priorities, constraints, and challenges.
- Attend strategic planning and roadmap meetings to understand the rationale behind prioritization decisions.
- Observe how leadership, market trends, and stakeholder demands influence decision-making.
- Use these meetings to ask questions and gain a deeper understanding of the broader business context. This reduces reliance on the PM as the sole "middleman" for business needs.
- Ask about budget, resource, and timeline constraints early in the design process to align expectations.
- Propose prioritization frameworks or methodologies (e.g., MoSCoW, Kano Model, or RICE scoring) to aid decision-making.
- Adjust design proposals to fit within these constraints while still striving for an excellent user experience.

2. Be Flexible with Design Iterations

- Be prepared to adapt your designs based on shifting priorities or constraints.
- Offer quick, low- or mid-fidelity prototypes that can be tested and validated with limited resources.
- Suggest scalable solutions that can be incrementally improved over time, rather than proposing large, complex overhauls.
- Break down ambitious ideas into smaller, actionable pieces that can be achieved incrementally.
- Focus on delivering Minimum Viable Product (MVP) features that can be expanded upon in future iterations.

3. Show Empathy and Support

- Just as you empathize with end users, extend that same understanding to your Product Manager.
- Acknowledge the pressures and constraints they face and appreciate the difficult trade-offs they must make.
- Offer your support by being a reliable problem-solving partner, demonstrating flexibility, and contributing ideas that balance user needs with business realities.

Be Proactive; Offer Solutions

Jump in, partner with me to
help solve problems you uncover.

As a Product Manager, I understand how frustrating it can be when design tasks arrive without the detailed user stories, acceptance criteria, or success metrics you need to do your best work. It's a common challenge, often stemming from the sheer demands of our roles, which can leave PMs stretched thin with little time to craft perfectly detailed design tasks. But instead of waiting for ideal conditions, I encourage you to be proactive.

Rather than waiting for a fully formed design task, consider joining me in the early stages of the process. Offer to participate in requirements-gathering, stakeholder meetings, or user interviews. By doing so, you'll gain firsthand insights that are often difficult to communicate through written tasks alone.

Your involvement at this stage helps you shape user stories and acceptance criteria, ensuring they're detailed, actionable, and aligned with both user needs and business goals. This collaborative approach not only enhances the quality of the design process but also reduces some of the workload on my shoulders, allowing us to move faster and more effectively.

> Routinely jumping in early will eventually result in creating more time for you to do design and will help transform how we work from being tasked with "Last Minute Visual Design" to more "Upfront UX" work.

Being proactive doesn't stop at design tasks. I encourage you to identify other opportunities where you can offer solutions to common challenges we face as a team. For example:

- If you notice conflicting requirements from stakeholders, propose a workshop to align priorities and find common ground.

- If resource limitations are slowing us down, suggest scalable design practices or tools that can streamline our workflows.

- When data is needed to support a decision, take the lead in conducting usability tests, user surveys, or competitive analyses to provide evidence-based insights. Don't wit to be asked, volunteer!

By stepping into these situations and addressing challenges head-on, you demonstrate your understanding of the broader business context and become a trusted partner, not just a recipient of tasks. Your proactive contributions build a stronger foundation for collaboration and ensure we're creating solutions that are both user-centric and strategically sound.

When you take this approach, the dynamic between us shifts from dependency to true partnership. Instead of feeling frustrated by incomplete tasks, you become an integral part of the problem-solving process, helping to shape our product from the ground up. This not only strengthens our working relationship but also ensures we're building products that balance user needs with business objectives. Together, we can create an efficient, collaborative, and innovative environment that benefits everyone involved.

Here are some practical steps to collaborate with PMs more proactively, align design with business priorities, and foster innovative, to create user-centric products that achieve shared success:

1. Collaborate on User Stories and Acceptance Criteria

- Work with Product Managers to co-write user stories, acceptance criteria, and success metrics.
- Create templates or guidelines that others can also use, to craft clear, actionable user stories.
- Incorporate your understanding of user needs to ensure stories are complete and focused.
- Keep acceptance criteria in mind when designing solutions to ensure designs align with business impact.
- Don't let the pursuit of perfection be an impediment to progress.

2. Preemptively Conduct Research

- Fill gaps in data or design reasoning by conducting usability tests, surveys, competitive analyses, or pattern research.
- Seek out user feedback, analytics, and support transcripts to identify problem areas.
- Use this data to cut through subjective opinions and focus discussions on evidence-based decisions.
- Don't wait to be asked; you may never be. Find ways to bring research results into the conversation, even if only a little at first.

3. Proactively Address Misalignment

- Recognize when priorities are misaligned and suggest alignment workshops or discussions to bring stakeholders together.
- If issues escalate, approach leadership with potential solutions, not just the problem.
- Facilitate collaboration by bridging gaps between teams and fostering shared understanding of goals.

#6
Speak My Language

*...and connect your design work
directly to business outcomes!*

Product Managers' primary goal is to ensure that the products we develop not only delight users but also align with our business objectives. To be truly impactful, UX Designers need to understand and speak the language of product management: the language of business.

It's not enough to simply understand the value of great design; you need to articulate its value in terms that resonate with business stakeholders. This means incorporating key performance indicators (KPIs), success metrics, and return on investment (ROI) into your thinking and storytelling.

When designers connect their work to business goals, it bridges the gap between creative problem-solving and strategic decision-making, enabling us to work together more effectively.

Here are three critical concepts every UX Designer should master to align their work with business objectives:

Key Performance Indicators (KPIs)

KPIs, or success metrics, are the quantifiable measurements we use to track how well our product is performing against business objectives. These metrics provide a clear benchmark for success and show whether our strategies are delivering the desired results. For UX Designers, understanding KPIs allows you to see how design decisions affect user behavior and how this ties directly to business outcomes.

For example, common KPIs include user retention rates, customer acquisition costs, conversion rates, and net promoter score (NPS). If a business goal is to reduce churn, focus on improving usability in areas where users are dropping off. A beautifully designed landing page is only valuable if it leads to conversion (sales). By tracking how your design decisions impact KPIs, you can make more informed choices and demonstrate the tangible value of your work to the business.

Use KPIs as a shared language between Design and Product Management. This helps us stay on the same page and work toward the same goals. It makes it easier to prioritize what really matters, connect design decisions to business impact, and show how your work drives results. Plus, it builds trust and makes collaboration smoother since you're both focused on hitting shared targets.

> A mindset focused on business outcomes will make you an indispensable partner to your PM and a more successful UX Designer.

Minimum Viable Product (MVP)

The concept of an MVP: a product with just enough features or value to satisfy early users and provide feedback for future development: is foundational to product management. For UX Designers, understanding MVPs is crucial because it informs design prioritization. Our goal is to deliver essential features that provide immediate value while leaving room for future updates and scalability.

The MVP approach doesn't just apply to an overall product and its initial launch or release. This approach can also apply to individual features or capabilities within a product. An MVP version of a new workflow can be released within an already established or legacy product. The first version of this workflow does not have to be fully researched, tested, perfect, or beautiful. There is often a lot of value in releasing something imperfect early, to gauge reactions, or understand how (or if) users will value it: then iterate based on user feedback.

Try to become comfortable with imperfection, design for iterations, embrace fast feedback loops, and in some cases, prioritize functionality over polish. By aligning with these MVP principles, you ensure that your designs are feasible for the short term and adaptable for long-term growth.

You can also explore the concept of a **Minimum Lovable Product (MLP)** to strike a balance between essential functionality and delightful design. This mindset helps both the business and users by ensuring that early releases are impactful without overcomplicating development.

Product Roadmap

The product roadmap is a high-level plan outlining the product's vision, direction, and key milestones over time. For UX Designers, understanding the roadmap ensures your designs align with both immediate priorities and long-term goals. By designing with the roadmap in mind, you can anticipate future features, avoid rework, and ensure that your work evolves as the product grows.

In other words, understanding the product roadmap allows UX Designers to better manage design scope and prioritize tasks effectively. By recognizing which features are critical for upcoming releases and which are planned for later stages, designers can allocate their time and resources wisely. This proactive approach helps balance delivering high-impact designs quickly while planning for more complex, long-term enhancements.

Collaborating with me to shape the product roadmap also provides opportunities for you to propose how and when we implement a phased approach to designing and delivering larger features. We can collaborate on which releases could include MVP versions of certain features, how and when we can gather feedback, conduct research, and form a plan for which follow-on releases could include incremental improvements in the user experience.

> When UX and Product are aligned on a roadmap, design becomes a strategic driver, contributing to both current and future business goals.

Alternately, without these things... without a Product Roadmap, KPIs, or Success Criteria, it is nearly impossible to collaborate towards a common end goal. It leads to reactionary work instead of strategic planning. It creates a frustrating environment without the ability to see incremental progress towards a shared outcome.

If you find yourself in a situation where the Success Criteria and/or a Product Roadmap are not well defined, lean in! Remind the team of their importance, ask questions, try to better understand the core problems the product aims to solve, explain how this can help everyone align, and gently push for some draft KPIs and an initial Roadmap: for the sake of the entire team.

Here are some other actionable steps Designers can take to align their efforts with business objectives, fostering collaboration with PMs to create impactful, user-centered products:

1. Study Key Performance Indicators (KPIs)

- Request and familiarize yourself with the KPIs most relevant to your product.
- Analyze past performance reports to understand how design impacts these metrics.
- Take a short course on data analytics or business metrics to strengthen your ability to measure design success.
- Make design decisions with KPIs in mind, and test how your enhancements impact them.
- Use KPIs as a shared language and common goals between you and your PM.

2. Learn About Minimum Viable Product (MVP) Approach

- Participate in MVP planning sessions (entire products or new features / capabilities) to understand how business priorities shape early product releases.
- Review case studies of successful MVPs in your industry to see how others balance delivering value with creating room for growth.
- Study lean startup methodologies to better grasp how MVPs tie into data-driven product iteration.
- Explore the concept of an MLP (Minimum Lovable Product) to balance functionality with delight.

3. Study the Product Roadmap

- Request access to the product roadmap and review it regularly.
- Attend roadmap planning and review sessions to understand the long-term vision and strategic goals.
- Collaborate with PMs to align your UX initiatives with the overall product and business strategies.
- Consider how your designs can scale or evolve to align with future features and milestones.

Get to Know the Business

...and really understand how
our products and UX drive it!

As a Product Manager, I've learned that balancing user experience with business objectives is crucial for creating successful products. While UX Designers excel at crafting visually appealing and user-friendly designs, it's essential to ensure those designs also support the broader business goals. This means stepping beyond the pixel-perfect interfaces and getting to know the business landscape in which the product operates.

To ensure our products meet user goals and business goals, we must look at UX Design through a business lens: How does user experience influence business metrics like retention, growth, and revenue.

To better understand how products drive the business (and how user experience can support) UX Designers can actively engage with other business units. UX Designers can benefit from attending leadership or strategy meetings to gain insights into high-level company goals. This could involve sitting down with marketing or sales teams to understand customer behavior and how it aligns with business priorities.

Let's take an example: Imagine working as a UX Designer at a large e-commerce company with a goal to increase sales by 15% over the next quarter. By attending leadership meetings, you might learn that abandoned carts are a major issue. Armed with this information, you could adjust your design focus to improve the checkout process. This might involve simplifying form fields, creating clearer calls to action, or adding features like persistent carts that follow users across devices. By addressing these pain points, your design aligns with the business goal of reducing cart abandonment and increasing sales.

In smaller organizations or startups, UX Designers often wear multiple hats, which gives them unique opportunities to influence product decisions more directly. This provides an opportunity to collaborate closely with PMs and other cross-functional teams to understand market positioning, revenue models, and competitive strategies. Attending startup pitch meetings, or even informal coffee chats with leadership, could provide a valuable window into how the company is trying to grow, what the biggest pain points are, and how great user experiences can drive the business forward.

As an example: As a designer at a fast-growing startup, your CEO might be laser-focused on scaling and gaining market share. With this context, you could sit down with the sales team to hear about the hurdles and challenges they're encountering in the market. If you learn that users are confused about how your product differentiates from competitors, you might rework the onboarding experience to highlight key value propositions. You might also add in-app tutorials or redesign feature discovery flows that educate users on the product's unique strengths. This way, you help the business capture more market share by directly addressing customer hesitations.

Stepping outside the bubble of pure design work and actively engaging with the business is one of the most powerful things a UX Designer can do. Whether you're in a large organization or a startup, aligning with business goals not only improves the value of your designs but also positions you as a key player in driving the company's success.

To truly integrate the larger business perspective into your design work, here are some practical actions you can take:

1. Attend Business Meetings and Briefings:

- Participate in product planning sessions, strategy briefings, and leadership updates. This keeps you informed about the company's goals, challenges, and technical constraints.
- The more you understand the bigger picture, the more effectively you can tailor your designs to support those objectives.
- Keep an eye out and listen for quarterly or yearly goals. Then consider how an improved user experience might help meet those goals.

2. Collaborate with Marketing and Sales Teams:

- Spend time with marketing or sales to understand customer behavior, buying patterns, and the metrics that drive revenue.
- Marketing and sales teams have direct access to customer feedback and market trends, which can inform your design decisions. They can help you see how your design affects sales conversions, customer satisfaction, or retention rates.
- Keep an eye out and listen for challenges and obstacles. Then consider how an improved user experience might help solve those challenges.

3. Analyze Competitors:

- Conduct competitive analysis (even informal) to understand market trends and where competitors are succeeding or failing in user experience.
- This analysis helps you find opportunities for differentiation. If competitors have a smoother onboarding flow or a more intuitive design, you can take those insights and iterate to offer a superior user experience.
- Look for competitors features and user experiences that set them apart. This gives you the chance to either keep up by matching those offerings, or innovate by designing a UI that's more intuitive or visually distinct, giving users a clear reason to choose your product over others.

#8
Let's Show Business
Value Together

...and collaborate to quantify the ROI of UX.

As a Product Manager, every decision I make must demonstrate clear business value. Whether it's choosing which features to prioritize, allocating resources, or setting project timelines, I am constantly under pressure to justify how each decision contributes to our overall business goals. This is especially challenging when it comes to justifying the value of good design. While we both know that great design enhances user experience, drives engagement, and differentiates our product in the market, it can be very difficult to actually quantify these benefits in terms that resonate with leadership.

For example, when I advocate for investing time and resources into refining the user interface or improving usability, I am often met with questions about the return on investment (ROI). Stakeholders typically want to see tangible metrics like increased revenue, higher user acquisition rates, or reduced churn. While the benefits of good design are real and significant, they are sometimes more subtle and long-term, making them harder to quantify in the short term. This is where your expertise and support as a UX Designer are invaluable.

To strengthen our case for the value of design, we need to present compelling, data-driven examples that link design improvements directly to business outcomes. This means providing research and case studies that show how design can improve key business metrics (KPIs), such as user retention, task completion rates, or Net Promoter Score (NPS). By grounding design decisions in data, we can demonstrate how intuitive, well-thought-out design directly impacts the bottom line. When we do this well, this leads to more time and funding to conduct more user research and design.

For example, research could show that simplifying navigation in a SaaS product can lead to increased task completion rates, which in turn boosts user satisfaction and reduces expensive customer support calls. Improved usability can lead to a higher NPS score, which correlates with customer loyalty and referrals: both of which drive revenue (and keeps us all employed!).

One of the most effective ways we can quantify the ROI of design is by collaborating closely to set measurable success metrics for each design initiative. Here's how we can do this together:

Define Success Metrics for Design

Start by aligning on the specific metrics that will be used to measure the success of a design change. This could include engagement metrics (e.g., time spent on a page), task completion rates, or customer satisfaction scores. For example, if the goal is to improve the onboarding process, we might look at metrics such as the time to complete onboarding, user retention after the first week, or the rate of users who return after onboarding.

Link Design to Business Metrics:

The next step is to tie these design metrics to broader business outcomes. For example, if we see that reducing the number of steps in the checkout flow leads to higher completion, we can link that to an increase in conversion rates and, ultimately, revenue growth. Similarly, if usability improvements in a feature reduce the number of user complaints, this can be tied to lower support costs and improved customer satisfaction.

Set Up A/B Testing or Experiments:

A/B testing is an effective way to measure the direct impact of design changes on user behavior and business metrics. We can test different versions of a design to see which one leads to better results, whether it's higher conversion rates, increased engagement, or improved retention. These experiments provide concrete data that we can use to make a compelling case to stakeholders.

For example, a product team might experiment with two different layouts for a landing page: one with simplified messaging and calls to action. The version that leads to higher sign-ups can directly demonstrate how design choices drive user acquisition.

If A/B tests are conducted solely with a lens on which design is best looking, without tying that user preference to a business outcome: a huge opportunity is missed to show ROI and make UX relevant to our leaders.

Monitor and Report the Impact:

After implementing design changes, track the performance of the chosen metrics over time. Documenting this data is crucial for showing the long-term impact of design on business goals. For instance, we could compare user retention rates before and after a major redesign to quantify how the improvements have positively impacted customer loyalty. If we see a 10% reduction in churn after improving usability, we can directly link that to business value, such as increased customer lifetime value (CLTV).

Continually measuring and reporting the quality, usability, and impact of your designs (even when they fail) builds the UX team's credibility as trusted sources of user research as well as everyones confidence in the product decisions that were made based on that research.

Here are some examples of metrics that work well in quantifying the ROI of UX design:

- **Conversion Rates:** Improved design of landing pages, forms, or checkout processes that reduce friction points can lead to higher conversion rates, which directly correlates with increased revenue.

- **Retention Rates:** Reducing churn by improving user experience, especially in critical areas like onboarding or troubleshooting, can lead to higher user retention, which boosts customer lifetime value (CLTV).

- **Net Promoter Score (NPS):** Intuitive design that delights users leads to higher NPS, which indicates customer loyalty and the likelihood of referrals: a key driver of organic growth.

- **Task Completion Rates:** Simplifying complex workflows or interfaces can lead to faster task completion, directly improving user satisfaction and productivity, which in turn leads to lower customer support costs.

- **Customer Support Volume:** Reducing customer confusion through better design can lead to a measurable decrease in expensive support calls, allowing the company to reallocate resources and reduce operational costs.

While PMs are committed to the value of good design, we need your help to articulate its business impact in a way that resonates with leadership and stakeholders.

By working together to gather evidence, present compelling arguments, and measure outcomes, we can ensure that design is recognized as indispensable to our product's success and secure further support and resources needed for further design and research.

As a UX Designer, you can play a critical role in helping to demonstrate the business value of good design. Here are a few actionable ways you can do that:

1. **Provide Data and Case Studies:**

   - Share case studies or research showing how design changes have led to improved business outcomes *in similar, even competing products*. For example, show how a competitor's product saw a 15% increase in conversions after simplifying their checkout process.
   - Leverage usability testing and user feedback to highlight how design changes address user pain points.

2. **Develop Success Metrics for Design Initiatives:**

   - Collaborate with your PM to define success metrics for design, such as user satisfaction scores, task completion rates, or engagement metrics.
   - Track and report on these metrics to show how design contributes to overall business goals.

3. Document Design Impact:

- Keep thorough documentation of design changes, user feedback, and the resulting business impacts. This can be shared with the broader team to ensure transparency and foster continuous learning.
- Present research findings that link design improvements to increased engagement, conversion rates, and customer satisfaction, using hard data to strengthen the argument for future design investment.
- Celebrate and evangelize successes where research and design led to measurable positive impact: this will foster further adoption of best practices.

4. Collaborate on A/B Testing and User Research:

- Set up A/B tests to compare the effectiveness of different design options, providing concrete evidence of the impact of design changes on key metrics like conversion rates or user retention.
- Run A/B Tests in Production *and* on Prototypes: both have value in conveying the value of good design to drive business outcomes.
- Use user research to identify pain points and areas of opportunity where design can make a significant impact, then quantify those impacts through testing and analysis.

#9
Provide On-Time Deliveries
...with Alternatives!

*Please try to deliver what I need,
when I need it ...and explain the risks.*

I understand that the nature of our work often puts us under tight deadlines to deliver new features swiftly. This sometimes results in requests for designs in a timeframe that doesn't allow you to fully think through, research, or test your ideas with end users. I know it can be frustrating to be asked to create design solutions with inadequate time, resources, or context to fully understand the user and the problem being solved. Unfortunately, sometimes that's just the nature of the business we are in.

Q: So how can UX Designers advocate for the time they need in a productive and collaborative way?

If you understand how a PM speaks (#6) and what the business needs are (#7 & #8), the answer is:

A: Communicate risks and manage expectations using a PM's language and potential impacts to the business.

One way to do this is for UX to borrow the classic adage, "*Good, Cheap, Fast: Pick Two...*" and adapt it in a way that works for the realities of product/design trade-offs.

To adapt the old adage to our needs, it can be reframed to focus on balancing time, quality, and confidence:

In our case, "Confidence" means: our comfort level that a proposed design will meet both user <u>and</u> business needs, because it has been validated with research and user testing.

Here is how it can be reframes for us:

"Good, Fast, Confidence: Pick Two..."

- **If it's Good and Fast, we won't have Confidence:** The design may look polished and meet core functional requirements, but without validation through user testing, we lack confidence that it solves the right problems for users <u>and</u> meets business goals. This comes with a *Low Confidence Score* and carries higher risks of needing rework later.

- **If it's Fast and Confident, it might not be Good:** A quick solution validated through basic testing may solve user needs but lack the rigor and polish necessary to fully meet business objectives or deliver a high-quality experience.

- **If it's Good and Confident, it won't be Fast:** Achieving both high quality and high confidence takes time. It takes time to explore multiple options, validate ideas, research and test with users, and refine down to the best possible solution. This approach ensures a *High Confidence Score* and minimizes the risk of failure post-launch.

When timelines are tight and speed is a priority, it's important to explain the trade-offs of delivering a low-confidence design. For example:

- *"If we prioritize speed, the solution may lack the depth of testing needed to confirm it meets user needs <u>and</u> business goals."*

- *"Design is about solving problems, not just making things look good. Without validation, we risk delivering something that works in theory but fails in practice."*

- *"Here's the trade-off: I can deliver a fast solution, but it may not be thoroughly tested or fully aligned with the problem we're solving. If we want higher confidence, we'll need to allocate time for testing and refinement."*

By framing the conversation around **Confidence**, we can emphasize that design is about achieving outcomes, not just delivering artifacts.

This helps Product Managers and stakeholders understand the risks of rushing, and opens the door to negotiating timelines or planning for iterative improvements post-launch.

Offer Alternatives

In addition, whenever you have low confidence that designs won't best meet user <u>and</u> business needs: offer alternatives! Maybe it's because the request was too prescriptive (*"We need a UI laid out this way…"*) or maybe the request was given without enough lead time (*"We need this in 2 days!"*). In either case: offer alternatives!

These alternatives don't always need to be polished. They can be low-fi designs, sketches, back-of-the-napkin drawings, or the results of a 6-Up exercise. The goal is to never stop reminding everyone that great solutions come from exploring alternatives (divergent thinking), then narrowing down (convergent thinking), and user testing (measuring impact).

Build Trust Over Time

We PMs love options! One of the best things you can do to partner with us is being open to creative ways to move forward by brainstorming options. Alternatively, one of the worst things you can do (that will certainly damper our progress and our relationship) is to hold fast to the idea that there is only one way forward, one option, especially if it's the expensive, lengthy option: even if it follows best practices and leads to the perfect design.

Instead, by repeatedly offering alternatives and routinely discussing trade-offs between Good, Fast, and Confidence, our PM/UX dynamic will evolve in a positive direction. At first, the "quick and dirty option" might be the popular, go-to choice. But, as we build trust and a track record of successful collaboration, we will likely start integrating more of your thoughtful design alternatives.

> With persistence, the long-term result will be: involving UX earlier in the process, more lead time, more time for research and testing, and more focus on increasing our confidence in what we develop and deliver.

By delivering what is needed on time and supplementing it with well-considered alternatives, discussing trade-offs, and their impact on KPIs, you can help bridge the gap between urgent demands and optimal user experiences.

Here are some actionable steps to get started implementing this approach:

1. Communicate Trade-Offs Clearly

- Use the "Good, Fast, Confident: Pick Two" framework to explain design trade-offs and the potential risks of prioritizing speed over quality or user insights.
- Celebrate examples where investing time in thoughtful design led to successful launches, better user adoption, and fewer revisions.
- Highlight specific examples of past instances where rushed designs led to unintended consequences, such as poor user adoption, or rework.
- Frame the conversation in terms of business impact, e.g., *"Without validation, we have Low Confidence this will improve KPIs like conversion rates."*

2. Offer Alternatives

- When faced with tight deadlines, present multiple solutions, such as a quick fix, a more informed design requiring additional time, or a phased approach.
- Provide low-fi options like sketches, wireframes, or simple prototypes to explore ideas without committing to full designs.
- Use divergent thinking techniques, like a "6-Up" exercise, to brainstorm alternative solutions and showcase the benefits of exploring multiple paths.

3. Advocate for Iteration

- Propose an iterative approach where a faster, initial design is delivered, followed by enhancements based on user feedback and testing.
- Highlight how incremental improvements can align with both business needs and user expectations over time.
- Track and share outcomes from iterative updates to demonstrate the value of taking a phased approach.

4. Build Trust Through Small Wins

- Deliver smaller, high-impact design solutions that address immediate needs while demonstrating the value of user-centered thinking.
- Use these quick wins as proof points to gain more time and resources for larger, more complex projects.
- Highlight how these wins directly contribute to meeting business goals: goals Product and Design should share: like improving user retention or reducing support tickets.

10

Offer Research and Reasoning

*If you see decisions being made without
data or reasoning, offer to help!*

Product Managers face constant pressure to deliver features quickly from stakeholders like Marketing, Legal, and executives. Strengthen our partnership and product quality by adopting a proactive, empathetic approach: lead with curiosity and offer to help.

When asked to design something without much explanation, background, or context, make it a habit to ask this key question:

"What research or reasoning do we have for heading in this direction?"

Maybe some research or reasoning already exists. If so, great: you'll get your hands on it and use it to guide your work. But maybe there isn't any research to back up the direction we've chosen. If that's the case, this is a gentle, empathetic way to highlight the fact that we're making a subjective or ill-informed decision.

After asking that key questions, if you get blank stares or no clear answers, that's great too! This creates a golden opportunity to inject design thinking and research. Leverage that opportunity to ask this follow up question:

"Can I help by collecting some data to guide our direction?"

If you get a chance to conduct or collect user research and the results support the decision that was already made, that's great! You're helping me and other decision-makers defend our direction. If the research provides a different perspective, that's good too: it's an opportunity to learn, redirect, and remind us to validate hypotheses in advance.

Asking these two key questions back-to-back isn't just about seeking clarity: it's about showing your commitment to, and reminding us all about data-driven decision-making.

> By demonstrating this curiosity, you position yourself as someone invested in the product's overall success: not just the design.

When these questions are asked repeatedly, over time, it helps change the culture to seek reasoning and validation upfront.

Believe me, when PMs say things like, *"We have to do this because Legal says so."* or *"This is what our investors want."* I probably share your frustration with these constraints. In those cases, please try to collaborate with me by delivering the designs we need on time. But also keep us all honest by asking, *"Can I help by collecting some data to guide our direction?"* Even if the immediate answer is *"No"*, your persistence in offering to gather data reminds us all to prioritize validation in the future.

Tailoring Research to Fit the Situation

When you get the chance to conduct research to inform our product or design direction, tailor your approach based on the type of decision we need to make. Once size doesn't fit all.

By specifying the type of research you're prepared to conduct, you make it easier for me to see its value, and choose an approach that fits the situation.

- **Usability Tests:** Quickly test a UI adjustment with a few users to validate changes and ensure they don't disrupt the user experience.

- **User Surveys:** Use short surveys to gather insights on user needs and preferences for strategic decisions.

- **Click-Stream Analytics:** Analyze existing navigation data to identify pain points or abandonment areas in user flows.

- **Support Feedback Analysis:** Review customer support transcripts to uncover common complaints or issues that inform product decisions.

- **Market Research:** Conduct quick competitor or market analysis to align decisions with industry trends and support leadership discussions.

When time is tight, prioritize methods that are quick and high impact:

- **Quick, High-Impact Methods:** Use usability tests, A/B testing, or analytics, existing logs, for fast, low-cost feedback.

- **Business-Focused Data:** Highlight metrics tied to business goals, like conversion rates or engagement, to support swift decisions.

- **Use Existing Data:** Tap into resources like Google Analytics, user feedback, or past research for immediate insights under tight deadlines.

Here are some practical steps to show empathy for the pressures PMs face, continually offering to help with data-driven insights, and improve collaboration through research …all to significantly enhance our collaboration and improve the product.

1. Ask About Existing Research and Reasoning

- When given a design request without much time or context, ask, *"What research or reasoning supports this direction?"* to encourage clarity and data-driven decisions.
- If no supporting data exists, offer to help gather insights by suggesting, *"Can I assist by collecting some data to guide us?"*

2. Focus on Business Goals & Leverage Existing Resources

- Align your research and reasoning with business priorities, using metrics like engagement, conversion rates, or revenue impact to communicate value effectively to PMs and stakeholders.
- Use existing data, such as customer feedback, analytics, or previous research, to quickly inform decisions and reduce time spent gathering new insights under tight deadlines.

3. Position Yourself as a Partner

- Demonstrate curiosity and a willingness to help by framing questions and research as collaborative efforts to strengthen the product.
- Show persistence in offering to gather data, even when immediate constraints prevent it. Your consistency fosters a culture of validation and reasoning over time.
- Use findings to help PMs defend their decisions or pivot as needed, emphasizing the shared goal of delivering the best product.

4. Encourage a Culture of Validation

- Regularly ask questions like, *"What can we validate before moving forward?"* to promote a habit of data-driven decisions.
- Share results from your research to show how it influenced better decisions, building trust and confidence in the value of UX contributions.
- Over time, help shift the team's mindset toward seeking research and reasoning upfront.

Most Important: Remember, your role goes beyond creating visually appealing designs: it's about partnering with your PM to ensure our product decisions are informed, impactful, and beneficial for our users and our business.

Conclusion

As we wrap up the list of things Product Managers wish you knew, one thing stands out: the power of partnership. By understanding the business, being supportive, and speaking the same language, you set the stage for an incredible collaboration.

When you bring curiosity and empathy into your relationship with your PM, you don't just improve your working dynamic: you create a foundation for designing products that truly resonate with users.

Remember, your Product Manager is often juggling a lot and under pressure to deliver. By showing them you understand their challenges, offer research, alternatives, and solutions, you're making their job easier. You're also demonstrating UX value. When you help everyone see the business value of your designs and meet deadlines with thoughtful alternatives, you're building trust and respect.

> This mutual understanding will, in turn, help ensure you are involved earlier, are given the lead time, resources, and opportunity to do more research and testing... so your creative ideas can become reality!

This book is one half of a larger conversation. For a PM and UX Designer to truly thrive together, both roles must embrace empathy, curiosity, and collaboration. This isn't about pointing fingers or delegating blame: it's about understanding and supporting each other to build great products together.

At the heart of it all, being a great partner is about empathy, communication, and a shared passion for solving problems. Focus on building a constantly evolving relationship centered on trust and respect. You'll find the process smoother and the outcomes more rewarding when you do.

So, keep that empathy flowing and remember, you're not just designing for users but also being an indispensable problem-solving partner in the pursuit of creating something truly amazing!

Bonus: A PM-UX Partnership Manifesto
10 Core Principles for Collaboration

What is a PM-UX Partnership Manifesto?

A manifesto is a shared set of principles aimed at strengthening collaboration between Product Managers (PMs) and UX Designers (UX). It outlines the values, practices, and mindsets that help both roles thrive together, ensuring that business goals and user needs are met harmoniously. Whether you're starting a new project or refining your existing processes, these principles serve as a practical guide to build trust, foster collaboration, and create better products. Use them in team discussions, roadmap planning, design critiques, or retrospectives to align your efforts and maximize your impact.

By embracing these principles, PMs and UX Designers can form a partnership that not only delivers exceptional products but also fosters a collaborative, innovative team culture. Together, we can create solutions that meet user needs, drive business success, and set a new higher standard for teamwork.

1. Shared Ownership of Product Success

We are united by a common goal: creating products that deliver value to users and drive business success. UX Designers and PMs share responsibility for achieving this outcome, with each bringing unique perspectives and skills to the table. Together, we solve problems, not just delegate tasks.

How to Apply: Start every project with a shared definition of success. Align on key metrics, user goals, and business objectives. Revisit this definition regularly to ensure focus.

2. Empathy for Users and Each Other

Empathy is at the heart of what we do: not just for our users, but for each other. PMs respect the time, expertise, and insights that UX Designers need to craft thoughtful solutions. UX Designers understand the constraints and pressures PMs face in balancing timelines, stakeholders, and goals.

How to Apply: Schedule regular check-ins to share challenges, insights, and updates. Use these moments to build mutual understanding and address roadblocks together.

3. Start with Problems, Not Solutions

We collaborate best when we frame our work around clear problems rather than predetermined solutions. By defining the "why" before the "what," we ensure that our efforts address meaningful user needs and business priorities.

How to Apply: Use problem-definition workshops, brainstorming sessions, or user journey mapping exercises to align on challenges before jumping into solutions.

4. Data-Driven Decision Making

We prioritize decisions grounded in data: user research, usability testing, and analytics: over assumptions or opinions. When data isn't available, we agree to test hypotheses and validate decisions through iterative processes.

How to Apply: Build time for research, prototyping, and testing into your roadmap. Use tools like confidence ratings or A/B tests to evaluate and refine ideas collaboratively.

5. Embrace Iteration and Exploration

The best solutions often emerge through exploration and iteration. We value the process of testing, learning, and refining over rushing to deliver the first idea. Multiple perspectives and feedback loops lead to better outcomes.

How to Apply: Allocate time for designers to explore multiple concepts and for PMs to review and discuss them. Plan iterative development cycles that include user feedback.

6. Transparent Communication

Clear, open, and proactive communication is the foundation of our partnership. We share context, goals, constraints, and progress openly to build trust and avoid surprises.

How to Apply: Use shared tools like project briefs, research repositories, or collaborative roadmaps to maintain visibility. Host regular syncs to align on priorities and expectations.

7. Respect Each Other's Expertise

PMs trust UX Designers to advocate for users and deliver high-quality experiences. UX Designers trust PMs to align the product vision with business goals and stakeholder needs. Each role values the other's expertise and contributions.

How to Apply: Involve each other early and often in discussions about priorities, strategy, and design. Acknowledge contributions and give credit for successes.

8. Focus on the End-to-End Experience

Great products consider the entire user journey, not just isolated features, or screens. We collaborate to design workflows and solutions that are seamless, intuitive, and aligned with the broader product vision.

How to Apply: Approach projects through the lens of user flows and systems thinking. Use tools like journey maps or experience maps to identify opportunities and gaps.

9. Advocate for Users, Advocate for Business Goals

Our partnership thrives when we strike the right balance between user needs and business objectives. UX Designers advocate for the user experience, while PMs ensure alignment with broader company goals. Together, we champion both sides.

How to Apply: Frame discussions around both user impact and business outcomes. Use user stories, metrics, and real-world examples to connect the two.

10. Foster a Culture of Collaboration

Collaboration isn't just a process: it's a mindset. We approach our work with curiosity, humility, and a commitment to learning from each other. We embrace feedback and view every interaction as an opportunity to strengthen our partnership.

How to Apply: Host team retrospectives to reflect on what's working and what's not. Celebrate wins together, no matter how small, and recognize the value of cross-functional contributions.

Practical Applications for Teams

- **In Kickoff Meetings**: Review the manifesto together to align on expectations and collaboration practices.

- **During Retrospectives**: Use the principles as a framework for evaluating team dynamics and identifying areas for improvement.

- **For Onboarding New Team Members**: Share the manifesto to introduce your team's approach to collaboration and the value of the PM-UX partnership.

- **As a Reference Tool**: Keep the manifesto visible: whether in a shared workspace or pinned in your collaboration tool: so it's always top of mind.

www.ingramcontent.com/pod-product-compliance
Lightning Source LLC
Chambersburg PA
CBHW071200210326
41597CB00016B/1609

Special Thanks to:

Brad Barnhart
Amy Battles
Cindy Brummer
Katie Cole
Kamala Espig

10 things your UX Designer wants you to know:

Introduction

Great Product Managers and UX Designers have one thing in common: they are both excellent problem solvers. They don't get distracted by things that don't matter. They stay focused on customer needs and how to solve them!

Yet, even with this shared mindset, PMs and UX Designers often struggle to work seamlessly together. Between them, there can be tension, pressure, conflict, disappointment, and unmet expectations. PMs often see UX Designers as overly focused on ideal solutions that look great but don't fully account for technical constraints, deadlines, or business priorities. Meanwhile, UX Designers sometimes feel PMs prioritize speed and business objectives over user experience, leading to rushed timelines and untested designs. UX Designers can feel frustrated when their input seems undervalued or when they're brought in too late to make meaningful contributions.

When great Product Managers and UX Designers collaborate effectively, they help everyone around them stay focused by continually asking the essential question: *"What problem are we trying to solve?"* So, when it comes to this book, let's take the same approach. Let's ask: What is the problem that keeps PMs and UX Designers from being great partners? Why do so many teams experience conflict, disappointment, or mismatched expectations? Is it a lack of process, skill, training, unclear roles, poor communication: or something else? Why do so many teams experience tension, frustration, or missed expectations?

Unfortunately, there isn't an overnight fix. There isn't a silver bullet. There isn't a single process change or intake/handoff mechanism that will magically create a great partnership between PMs and UX Designers. Every team is different, and the best collaborations take time and effort to develop.

So, what can a PM do to make the most of their collaboration with UX? Fortunately, it lies in leveraging strengths you already have! One of the greatest strengths of Product Managers is their ability to balance **strategic thinking** with **leading multi-disciplined teams**. You understand how to prioritize what matters most: juggling the needs of sales, marketing, engineering, design, and other stakeholders: while keeping the team aligned toward a shared goal. Your skill in synthesizing diverse inputs and focusing on the bigger picture makes you an indispensable leader in the product development process.

This strength is invaluable when it comes to your partnership with UX Designers. By sharing a clear understanding of priorities and providing context on business goals, market demands, and technical constraints, you help UX Designers channel their creativity into solutions that solve the

right problems. At the same time, your ability to unite and inspire cross-functional teams ensures that every discipline, from engineering to design, contributes their best work in pursuit of a cohesive vision.

When you combine **strategic prioritization** with **collaborative leadership**, you unlock the full potential of your team: including UX Designers. By aligning on goals, providing clarity, and fostering mutual respect, you create a partnership that delivers products that are not only innovative and user-centered but also aligned with the business's overarching objectives. You can transform complexity into clarity and deliver outcomes that resonate with both users and stakeholders.

This book isn't about placing responsibility solely on you the PM or your UX Designer to fix it. It's about working together, leveraging each other's strengths, and building trust. When PMs and UX Designers collaborate effectively, they not only create better products, but they also find greater satisfaction in their work: and more success in their careers.

This book is one half of a larger conversation. For a PM and UX Designer to truly thrive together, both roles must embrace empathy, curiosity, and

collaboration. This isn't about pointing fingers or delegating blame: it's about understanding and supporting each other to build great products together.

With that in mind, here are ten things UX Designers want you to know, *in their words*.

#1
UX Is More Than Design.
It's Problem-Solving.

*Unlocking the full, strategic
value of UX, beyond aesthetics.*

As a UX Designer, I understand the pressure you face as a Product Manager. You're constantly balancing business goals, stakeholder demands, tight deadlines, and technical constraints. Your ability to prioritize, pivot, and drive progress in the face of so many competing priorities is impressive. I see how much responsibility you carry in aligning teams with all kinds of varying skillsets and coordinating them to deliver results.

That's why I want to partner with you: not just as someone who designs screens, but as a strategic collaborator who can help solve complex problems. UX is often misunderstood as simply being about how things *look*. But effective UX design is about how things *work*. It's about deeply understanding user behavior, identifying friction points, and creating experiences that are intuitive, engaging, and aligned with business goals.

When UX is treated as an afterthought, or as a group that "makes things pretty", we miss significant opportunities to create solutions that address real user and business problems. When UX is brought in early, given the problem to solve, is able to partner with the rest of the team, and is

included in strategic conversations, we can better help create solutions that not only delight users but also deliver measurable business impact.

UX Is More Than Visual Design

When many people hear the word design, they immediately think of colors, fonts, and layouts. While this visual design is important, UX design goes far beyond aesthetics. It's about understanding how people interact with products and how those products solve user problems. UX design is deeply rooted in research, testing, problem-solving, and creating intuitive workflows that make users' lives easier.

If I am only brought in after a problem has been solved, after a solution has been determined, then asked to simply make that solution look good, my contribution will be limited: and may not be helpful. In those scenarios, I become a decorator, not a problem-solver. But when I'm involved earlier: before decisions are finalized: I can help shape solutions that align with both user needs <u>and</u> business objectives.

UX design isn't about adding polish at the end; it's about working collaboratively to identify problems upfront, test assumptions, and design

products that deliver value. When we approach design as a method of problem-solving, the results are far more effective: for both users and the business.

The Strategic Power of UX in Problem-Solving

Imagine we're redesigning a checkout flow for an e-commerce platform. If I'm only asked to make the page more visually appealing, I might choose better colors or fonts, but that may not have a big positive effect on the overall outcome. If I'm brought in early to explore why users abandon their carts, I can conduct user research, analyze behavior patterns, and uncover deeper issues: like confusing shipping options or unclear pricing.

By solving *those* problems, simplifying the workflow, or rethinking how we explain pricing, we can reduce cart abandonment and increase conversions. That's the true power of UX: not in visual enhancements, but in solving the right problems with user-centered solutions that drive business results.

Another example might be customer onboarding for a SaaS product. A PM might ask for a sleeker, "more modern" onboarding screen, but early UX

involvement could reveal that users struggle with understanding how to get started. This insight might lead to rethinking the onboarding steps, which features to explain in what order, offering tooltips, or providing guided tours: resulting in higher user activation and customer retention.

These examples highlight how UX can help uncover root problems and create solutions that make a real impact. That is why early collaboration between us is critical.

To fully tap into the potential of UX, it's essential to think of it as a strategic, problem-solving discipline, not an artistic endeavor.

How This Approach Benefits the Product

- <u>Better Problem Definition</u>: We can use research, user interviews, and data to make sure we're solving the right issue.
- <u>More Effective Solutions</u>: Brainstorming together helps align user needs with business goals.
- <u>Faster Iteration</u>: Early testing helps avoid expensive pivots later in the process.
- <u>Reduces Rework</u>: Solving the right problems early avoids costly redesigns later in development.
- <u>Drives Business Outcomes</u>: Well-researched, user-centered solutions directly impact KPIs like retention, engagement, and conversion rates.
- <u>Strengthens Team Alignment</u>: Focusing on problems creates a shared purpose, aligning UX, PMs, engineering, and stakeholders.
- <u>Unlocks Innovation</u>: Problem-solving opens space for creative, strategic thinking beyond surface-level fixes.

When UX and PM Work Together, Great Things Happen

A great partnership is created when a UX Designer leans on the PM for direction and priority, the PM understands the value of UX goes beyond visuals, and they work together to create solutions to problems. Instead of handing off feature specs, we can worked together from the start to map out user pain points, sketch early concepts, and test rough prototypes with users. We can avoid unnecessary development work, refine our approach based on real feedback, and launch a feature that both users and stakeholders love.

A Stronger Partnership = Better Products

At the end of the day, we both have the same goal: building products that solve real problems, delight users, and achieve business goals. When we collaborate closely and leverage each other's strengths, we can create better experiences that not only look good but actually work well.

Let's shift our company's mindset from "UX makes it pretty" to "UX helps solve problems." By bringing UX in earlier and using UX strategically, you'll unlock more value and create products that have a bigger impact.

So, how do you get more out of your UX team? Here are some things PMs can do to get the most out of UX Designers by leveraging their skills more effectively:

1. **Engage UX in Root Cause Analysis:**

   - Include UX in user feedback reviews, customer support discussions, or data analysis sessions.
   - Leverage UX to uncover the root causes behind low adoption or feature underperformance.
   - Work together to propose user-centered solutions based on real insights.

2. **Ask for Insights, Not Just Designs:**

   - Remember, we're trained to think about user psychology, cognitive biases, and behavioral patterns.
   - Tap into user insights when planning features, roadmaps, and success metrics.
   - Use UX findings to inform business decisions, not just product visuals.

3. Use UX for Prioritization:

- Not all user problems are equal; UX can help identify which issues are most critical to fix.
- Leverage UX research to determine the highest-impact areas for development.
- Use qualitative and quantitative data to back up prioritization decisions.

4. Think Beyond UI:

- UX covers everything from user flows and information architecture to accessibility and emotional responses.
- Leverage UX skills to create holistic experiences, not just visual ones.
- Ensure UX considerations extend into content strategy, micro-interactions, and onboarding experiences.

#2
Better Results Come from Early Collaboration

Early UX involvement leads to smarter solutions.

As a UX Designer, I understand how much is on your plate as a Product Manager. You're balancing business goals, stakeholder expectations, development timelines, and countless decisions that shape our product's future. I respect how much pressure you're under to deliver outcomes efficiently and effectively. That's why I want to work *with* you: not just *beside* you: to ensure we're building the right solutions, not just delivering features.

One of the most effective ways we can create meaningful, impactful products is by collaborating early in the product lifecycle. Involving UX early isn't just about having more time to make designs look better: it's about solving the right problems before they become costly to fix. When UX is brought in too late, we're often tasked with polishing something that doesn't fully meet user needs or align with business goals. But when we collaborate from the start, we can shape solutions that are user-centered, feasible, and valuable for the business.

Early UX Involvement Reduces Risk and Rework

Imagine we're building a new feature, and by the time I'm looped in, key decisions about how it should work have already been made. At this point,

my role becomes limited to tweaking visuals or making small usability adjustments. This reactive approach might get the feature out the door faster, but it risks missing the mark with customers. Worse, it could result in costly rework down the line if users struggle with it, or abandon it altogether: causing us to revisit it, redevelop it, and re-release it.

Now, imagine a different scenario. I'm brought into the conversation early: before features are fully defined. Together, we explore user pain points, brainstorm solutions, and test concepts before development begins. In this scenario, we can spot risks early, validate assumptions, and align the user experience with our product strategy. This proactive approach doesn't slow us down; it helps us *move faster in the long run* by reducing costly pivots later.

> Involving UX *early* is a **risk-reduction strategy**, not a luxury. Early collaboration ensures we build the right thing the first time, saving time, money, and effort.

Designing with the Product Vision in Mind

When UX is included in early discussions about product vision and roadmap planning, we can design with scalability and long-term goals in mind. This foresight allows us to anticipate how features will evolve and how user flows will expand as the product grows. Without that context, design decisions can become shortsighted, leading to fragmented experiences that need to be reworked as the product scales.

For example, let's say we're introducing a new onboarding flow. If I understand the product's future plans: such as upcoming features or new user segments: I can design an onboarding experience that *grows and scales with the product*. This foresight avoids the need to overhaul the design later when new requirements emerge. Early alignment allows us to solve today's problems while preparing for tomorrow's opportunities.

Co-Creating Solutions Leads to Better Outcomes

The best solutions often emerge when teams co-create. That's why I value opportunities to work alongside you in defining problems and brainstorming solutions. When UX is part of stakeholder interviews, and

problem-definition workshops, we gain firsthand insight into the challenges we need to solve. This collaborative process leads to more innovative solutions that balance user needs with business goals.

Consider this: Instead of handing me a feature spec that says, "*Design a new settings page*", let's explore the user problems that make this feature necessary. What tasks are users struggling to complete? How does this feature fit into their broader workflow? What business outcomes are we trying to drive? Answering these questions together allows us to design solutions that truly address user pain points and align with product objectives.

This doesn't mean slowing down the process or holding endless meetings!

It's about creating space for quick alignment: whether that's a brainstorming session, a user journey mapping exercise, or reviewing research findings together. These touchpoints help us stay on the same page and make more informed decisions.

Early Testing Saves Time and Builds Confidence

Another advantage of early collaboration is the ability to test ideas before they're locked in. When I have time to explore multiple concepts and validate them with users, we can catch usability issues and misalignments before development begins. This reduces the risk of launching features that don't meet user expectations or require expensive fixes after release.

For example, if we're designing a dashboard, I could create several low- or mid-fidelity prototypes that highlight different ways to prioritize user tasks. We could then quickly test these concepts with users to see which approach works best. This kind of early feedback loop allows us to refine solutions before committing engineering resources. It's faster and cheaper to test and refine designs than to rebuild features after launch.

A Shared Commitment to Building the Best Product

At the end of the day, we both want the same thing: to build products that succeed. Early collaboration between UX and Product Management isn't about adding extra steps: it's about working smarter together. By aligning

early, we can move faster, reduce risk, and create products that are not only functional but also delightful and impactful.

This partnership works best when it's built on mutual trust and shared ownership of the product's success. I'm not just here to make things look good: I'm here to help solve problems, drive business results, and create experiences that users love. The earlier we collaborate, the stronger and more strategic our solutions will be.

Here are some things we can do together to takes some steps towards a better partnership resulting in better products:

1. **Involve UX in Discovery**

    - Instead of waiting until wireframes or prototypes are needed, bring UX in when we've first identifying a business problem.
    - Let user research, insights, and data help shape the strategy.
    - Collaborate with UX to define success metrics from the start.

2. **Include UX in Planning**

    - Invite UX Designers to brainstorming sessions, and early-stage planning meetings.
    - Share any user feedback, previous user research findings, or insights, to allow UX to design with those in mind.

3. **Encourage Co-Creation of User Stories**

    - Collaborate with UX when defining user stories and requirements.
    - Focus on framing problems, not prescribing solutions, to encourage user-centered designs.

#3
Sharing Goals Helps Us Design for Impact

Aligning on desired outcomes leads to better results.

As a UX Designer, my role is to create experiences that solve user problems and make our product intuitive and delightful. But great design doesn't live in isolation: it thrives when it's connected to the broader business context. When I understand the product goals, the market pressures, and how success is measured, I can design with purpose and impact. That's why sharing goals and success metrics with me isn't just helpful: it's essential for creating designs that move the needle for both users and the business.

Why Shared Goals Matter

You, as the Product Manager, are the strategic driver of the product. You have deep insights into business goals, stakeholder expectations, and market dynamics. I respect that you're constantly balancing user needs with business realities. The challenge is: sometimes those business drivers don't always trickle down to the design process. Without visibility into the "why" behind product decisions, my design solutions may miss opportunities to have a bigger impact or contribute to larger outcomes.

By openly sharing your goals and the key performance indicators (KPIs) that matter, you empower me to design solutions that not only work for users but also support our business objectives. This alignment turns design from a support function into a strategic driver of product success.

> When UX and Product align on goals and success metrics, design becomes a powerful tool for driving measurable business results.

Designing with Purpose

Without understanding the product's goals, I might focus on solving the wrong problems or optimizing areas that won't actually help the business. For example, I could spend time perfecting a feature's aesthetics when the real need is to reduce user drop-off during checkout. However, if I know that increasing conversion rates by 5% is a top priority, I can focus my design efforts on streamlining workflows and reducing friction where it matters most.

Consider a scenario where we're tasked to redesign a product's home screen. If I know that one of our core KPIs is improving user engagement

with key features, I can prioritize making those features more discoverable and accessible. On the other hand, if our primary goal is reducing churn, I might focus on simplifying complex workflows or enhancing onboarding to help users find value faster.

When goals are clear, design decisions become more intentional. Every color choice, layout, and interaction can be justified because it serves a shared objective. This not only creates a better user experience but also delivers tangible results for the business.

Understanding the "Why" Behind Product Decisions

I understand that as a PM, you're often making decisions based on a variety of factors: customer feedback, market analysis, technical constraints, and business strategy. But sometimes, to the design team, decisions can seem arbitrary or disconnected from user needs. This isn't a reflection of your intent; it's a communication gap.

When you take the time to explain why certain features are prioritized or why a deadline has been accelerated, it provides valuable context.

Knowing that we need to launch a feature quickly to capitalize on a market trend or respond to competitor's latest release helps me make smarter design trade-offs. I can focus on delivering a functional solution now while planning enhancements for later.

Similarly, if leadership is pushing for a feature because it's tied to a new revenue stream, sharing that reasoning allows me to consider how the design can drive adoption and engagement. Context turns constraints into opportunities and helps me design with the bigger picture in mind.

Connecting UX Work to Business Impact

One of the most effective ways to demonstrate the value of design is by linking it to business outcomes. When UX efforts directly contribute to KPIs like conversion rates, retention, or customer satisfaction, it becomes easier to advocate for design resources and influence product strategy. For example, if simplifying a form leads to a 20% increase in sign-ups, or improving onboarding reduces churn by 15%, that's not just a design win: it's a business win. When we measure and communicate these impacts, it builds a stronger case for continued investment in UX.

To achieve this, I need your partnership in defining what success looks like. Which metrics are most important for this release? Are we focused on acquisition, engagement, or retention? Should we prioritize customer satisfaction (CSAT), Net Promoter Score (NPS), or feature adoption rates? Answering these questions together ensures that design work is not only user-centered but also aligned with strategic business goals.

Designing for Short-Term Wins and Long-Term Growth

Business goals often operate on multiple horizons: some are immediate, like hitting quarterly revenue targets, while others are long-term, like building brand loyalty. Knowing how these timelines affect our product strategy helps me prioritize design work.

If we need to launch a feature quickly to meet a short-term goal, I can focus on delivering a minimum viable product (MVP) with plans for iterative improvements. Conversely, if we're investing in a long-term initiative, I can dedicate more time to research and design exploration, ensuring the solution is scalable and future-proof.

Balancing these priorities requires open communication and a shared understanding of what's urgent and what's important.

Here are some things we can do to get aligned, so we can deliver short-term wins without sacrificing our long-term vision:

1. **Communicate Business Drivers**

   - Share the "why" behind product decisions to give UX designers meaningful context.
   - Discuss how market conditions, stakeholder goals, or competitive pressures shape priorities.

2. **Establish a Shared Metrics Framework:**

   - Work with UX to define a set of shared KPIs that balance business objectives with user experience goals.
   - Ensure that both qualitative (e.g., usability feedback) and quantitative (e.g., conversion rates) metrics are considered.
   - Add a dedicated KPI field in tracking tools like Jira to ensure each design task is explicitly linked to a relevant success metric.
   - Regularly revisit and refine the framework to adapt to evolving product goals and measure design impact effectively.

3. **Integrate UX into Routine Product Reviews:**

    - Involve UX Designers in quarterly or monthly product reviews where KPIs are discussed and analyzed.
    - Provide access to product performance dashboards so UX can track how design efforts influence key metrics.

4. **Align Design Sprints with Business Objectives:**

    - At the start of each design sprint, clearly outline how UX work can contribute to specific business goals and success metrics.
    - Ensure each design task in Jira includes a field that links the task to a defined KPI, helping to track and validate its impact.
    - Prioritize design initiatives based on their potential to influence critical success metrics such as retention, engagement, or CSAT.
    - Incorporate sprint check-ins to review progress against KPIs and adjust priorities as needed.

#4
Let's Solve User Problems Together
...focusing on real problems, not screens or features.

As a UX Designer, my goal is to create experiences that make users' lives easier, more enjoyable, and more efficient. But to design truly impactful solutions, I need to deeply understand the problems users are facing: not just be handed a list of features to design. When our collaboration focuses on solving real user problems rather than simply building features, we create products that deliver meaningful value to both users and the business.

I understand the pressure you face as a Product Manager. You're balancing stakeholder demands, development timelines, and business objectives. Often, this means you're being asked to deliver specific features by a certain deadline. But when we start with a predetermined feature in mind, instead of a clearly defined user problem, we risk building something that may not actually address users' needs. Together, we can improve that.

> The most successful products solve real user problems. When we focus on problems: not just features: we deliver *lasting value*.

Shifting the Focus: From Features to Problems

It's easy to fall into the habit of framing work around features. Stakeholders often come to you with requests like, *"We need a new dashboard"* or *"Let's add a search filter."* But behind every feature request is usually an underlying user problem. What if, instead of immediately jumping in to design the feature, we paused to ask:

- What problem is the user experiencing?

- Why do they need this feature?

- How will this solution fit into their workflow or journey?

For example, imagine a request comes in for a new reporting feature. Instead of jumping into wireframes, we could explore:

- Are users struggling to find the data they need?

- Is the current reporting process too complex or time-consuming?

- Are users looking for ways to share insights with others?

By understanding the problem, we open the door to solutions that may be more effective than the original feature request: maybe even faster and less costly to implement. Maybe the answer isn't a new reporting feature: it could be simplifying existing data views, providing export options, or automating report generation. This problem-first approach leads to more innovative, user-centered solutions.

User Research to Lead the Way

Understanding user problems requires research. This can include user interviews, usability testing, surveys, or analyzing support tickets and product data. When UX is involved early in the discovery process, we can uncover pain points that might not be obvious.

User research doesn't have to be lengthy or expensive. Even quick conversations with a handful of users or lightweight testing of prototypes can reveal insights that guide better product decisions. When you include and prioritize UX-led research, you empower us all to bring evidence-based insights to the table.

For instance, if we're hearing requests for more customization options in a product, user research might reveal that users aren't necessarily asking for more settings: they're struggling to complete tasks because the defaults aren't well-optimized. In that case, improving the default experience could be more impactful (and less costly) than building an elaborate customization system.

Research uncovers the "why" behind user behaviors, helping us design solutions that address root causes rather than surface symptoms.

Looking Beyond Features: The Full User Journey

Users don't interact with products in isolated screens or features: they experience products as complete journeys. By thinking beyond individual features and looking at the entire user workflow, we can identify opportunities to improve the holistic experience.

Let's say we're tasked with redesigning a settings page. On the surface, this might seem like a straightforward UI update, requiring just refresh to the "look and feel". But if we zoom out and examine the full journey, we might discover that users only visit settings because they're confused

elsewhere in the product. Improving the clarity of navigation or default behaviors could eliminate the need to visit the settings page altogether.

When we can, we should explore: What is the customer trying to accomplish? Why are there here? How did they get here? When do they come here? How often? Where do they go next? Where do we want them to go? This approach leads to more cohesive and intuitive designs that better meet the needs of our users and align with the product's vision.

By encouraging end-to-end thinking, we can:

- Identify bottlenecks in the user flow.

- Reduce friction between steps in a process.

- Design for outcomes, not just interfaces.

When we consider how a feature fits into the broader user journey, we can design solutions that feel seamless and intuitive.

Building a Culture of Problem-Solving

It takes time to create a product culture that prioritizes solving user problems, but it starts with asking the right questions. Together, we can build a practice of pausing to define the problem before jumping into solutions.

Let's foster a culture of curiosity by:

- Beginning product discussions with the question: *"What problem are we solving?"*

- Validating assumptions with user research before defining solutions.

- Encouraging brainstorming sessions focused on user pain points rather than feature lists.

By being curious and working together to identify and solve real user problems, we can create products that are not only functional but also meaningful and impactful. Let's make problem-solving our shared priority!

Here are other some things Product Managers can do to help us all focus on problem solving, so we can not only not only create better products but also strengthen our partnership and collaboration:

1. Frame Features Around User Problems

- Start product conversations with, "*What problem are we solving*?" rather than jumping to solutions.
- Encourage teams to dig into user pain points and workflows before defining features.

2. Let User Research Lead the Way

- Prioritize and support UX-led research to uncover user needs and pain points.
- Involve UX in discovery phases and decision-making processes to ensure solutions are evidence-based.

3. Encourage End-to-End Thinking

- Collaborate with UX to understand how features fit into the entire user journey.
- Focus on designing workflows and experiences, not just individual screens or interactions.

4. Bring in Multiple Perspectives

- Connect UX with cross-functional teams (Marketing, Sales, Customer Support) to share diverse insights into user problems.
- Support UX-led A/B testing and experimentation to validate design decisions with real user data.

#5
Let's Shape Our
Product Roadmap Together

UX can help prioritize what truly matters most for success.

I understand that being a Product Manager comes with a unique set of pressures. You're constantly balancing the demands of various stakeholders: whether it's sales pushing for new features, marketing aiming to stay ahead of competitors, legal ensuring compliance, or engineering considering technical constraints.

The entire team relies on your ability to identify and prioritize the right problems to solve, ensuring that we're focusing our efforts where they will have the most impact. Your perspective is critical because you see the product through multiple lenses, managing risks and aligning competing interests. These decisions ultimately shape the product roadmap, which serves as our guiding strategy. As a UX Designer, I can help by ensuring that user insights and experience are considered and integrated into the roadmap, helping us build a product that not only meets business goals but also delivers meaningful value to our users.

UX as a Strategic Partner in Roadmap Planning

By shaping our roadmap around users' top pain points from user research and usability testing, we can prioritize high-impact features that deliver value and support long-term business success. Together, we can make the roadmap more strategic, user-centered, and impactful.

A well-defined product roadmap benefits the entire team: it provides clarity, aligns efforts across departments, and creates a shared understanding of where we're headed. It helps avoid reactive decision-making and instead allows us to proactively steer the product toward long-term goals.

Being Proactive About the Product Vision

As the Product Manager, your role isn't just to react to requests from sales, marketing, legal, and stakeholders; it's to proactively shape the product vision. A roadmap should not be a collection of feature requests, but a strategic tool that aligns with our long-term vision. By setting clear

priorities based on real user insights and business goals, we can ensure we're not just building features but solving meaningful problems.

Similarly, you can ask UX to help shape the long-term direction of the product, think beyond immediate needs, and create a **UX Vision**. This doesn't have to be a fully researched, tested, or functional prototype. It could be just a few screens or slides of the main "happy path" workflow to paint a picture of *what could be*. Even if the UX vision isn't entirely achievable in the short term, it sets the tone and inspires the entire team to think about the ideal user experience we're striving for. This forward-thinking approach helps align efforts and ensures we don't lose sight of what truly makes our product valuable to users.

Grounding the Roadmap in User Research and Testing

User research provides a clear, evidence-based understanding of what users need most. These insights guide feature prioritization, ensuring that resources are allocated where they will have the greatest impact. Instead of relying on assumptions or stakeholder requests, we can make informed decisions rooted in real user behavior and needs.

Usability testing plays a critical role in shaping the roadmap by identifying friction points and opportunities for improvement early in the process. Testing prototypes or early designs allows us to catch potential issues and refine solutions before they reach development.

For instance, if usability testing reveals that users struggle with a new navigation system, it signals the need to prioritize the roadmap around improvements in findability and information architecture. Addressing these issues upfront helps prevent costly rework and ensures that future iterations focus on refining the user experience. Regular user testing helps validate assumptions, prioritize enhancements based on user needs, and guide the roadmap toward a more intuitive and successful product.

Balancing Short-Term Business Goals & Long-Term User Satisfaction

There will inevitably be times when user-focused recommendations don't align with immediate business goals. In these cases, I can advocate for long-term user satisfaction by framing features in terms of their broader business value.

For example, if a feature doesn't directly drive short-term revenue, we might highlight how it enhances customer retention, reduces support costs, or builds user loyalty: outcomes that contribute to long-term success.

As a UX Designer, when advocating for user-centered features, I focus on:

- **Connecting User Needs to Business Objectives:** Showing how addressing user pain points supports metrics like retention or engagement.

- **Proposing Data-Driven Experiments:** Suggesting A/B tests or pilots to validate a feature's value with minimal risk.

- **Identifying Quick Wins:** Highlighting low-effort, high-impact features that improve satisfaction and build momentum for the roadmap.

Here are some things we can do together to ensure the roadmap reflects what matters most to users and stakeholders, leading to a product that's successful, well-loved, and impactful:

1. Establish a User-Centered Roadmap Structure

- Create Themes Based on User Needs: Group roadmap items into themes aligned with user journeys or needs, such as "Improving Onboarding" or "Streamlining Workflows," rather than technical or business categories.
- Tie Roadmap Items to KPIs: Ensure that every feature or design task is linked to a key performance indicator (KPI) in tracking tools like Jira to track progress and impact effectively.
- Define a UX Vision for the Future: Encourage UX to think beyond immediate needs and craft a long-term vision that helps shape product decisions and align the team toward a cohesive experience.

2. Ensure Continuous Alignment Through Review and Adaptation

- Hold Regular Roadmap Review Meetings: Schedule dedicated meetings (monthly or quarterly) to assess how features align with user research, evolving business needs, and usability findings.
- Adjust Priorities Based on New Insights: Keep the roadmap flexible and responsive by incorporating feedback and adapting to emerging trends and data.
- Encourage Cross-Functional Participation: Involve stakeholders from product, engineering, and UX to ensure alignment and collective ownership of the roadmap.

3. Prioritize Features That Maximize User and Business Impact

- Evaluate User Impact: Assess each feature's potential to address user pain points and enhance satisfaction, ensuring value-driven prioritization.
- Identify Quick Wins: Focus on low-effort, high-impact improvements that can provide immediate value and build momentum.
- Balance Short-Term and Long-Term Goals: Align feature prioritization with both immediate business needs and long-term product strategy for sustainable growth.

4. Integrate UX Testing and Validation into Roadmap Execution

- Allocate Time for Usability Testing: Build in dedicated testing phases to validate assumptions, iterate on designs, and refine features before full development.
- Incorporate Feedback Loops: Use findings from usability testing and customer feedback to adjust and improve designs iteratively.
- Minimize Risk Through Early Validation: Catch potential usability issues early to avoid costly changes and ensure a smoother user experience at launch.

#6
Defending Product Decisions as a Team

How UX and PMs can collaborate to build stakeholder confidence.

As a UX Designer, I understand the immense pressure you face to justify product decisions to stakeholders. With competing opinions, tight deadlines, and business objectives, explaining why certain features are prioritized over others can be challenging. I'm here to help by grounding our decisions in user research, data, and feedback. Together, we can build a united front to not only defend our choices but also demonstrate how they address real user problems and align with business goals.

Grounding Decisions in Research and Data

Defending product decisions becomes significantly easier when they're backed by evidence. UX research provides tangible insights into user behavior, pain points, and needs. By conducting usability tests, user interviews, and analyzing analytics, we can validate our decisions and demonstrate their impact on user satisfaction and business metrics.

For instance, imagine we're building a new e-commerce purchase flow for a subscription-based app. Our research shows that 60% of users drop off at the payment step due to unclear instructions. Based on these insights, we prioritize redesigning the payment page to include step-by-step guidance and error prevention features. When presenting this decision to

stakeholders, tying it to potential improvements in conversion rates reinforces the value of this prioritization. Stakeholders can clearly see how addressing this pain point translates into increased revenue and happier customers.

Presenting Unified Research Narratives

One of the most effective ways to defend product decisions is by presenting a cohesive narrative that connects user insights to business objectives. This means telling a story that walks stakeholders through our team's journey from problem identification to proposed solution, backed by data at each step.

For example, let's say stakeholders push to deprioritize a feature aimed at improving accessibility in favor of a flashier, marketing-driven update. Together, we can present research showing that 20% of our user base struggles with accessibility issues, leading to poor task completion rates and higher support costs. By connecting this data to business goals like customer retention and cost savings, we can explain why improving accessibility aligns more closely with long-term success. This narrative

helps stakeholders see the logic behind our decisions and shifts the conversation from subjective opinions to data-driven insights.

Co-Presenting to Build Stakeholder Confidence

Stakeholders are more likely to trust decisions when they see alignment between UX and PM. Co-presenting research findings and product decisions demonstrates that we're on the same page and have considered every angle.

Consider a scenario where a stakeholder questions why a particular feature was delayed. During the presentation, you can explain how competing priorities from different departments required a reevaluation of timelines, while I share user feedback showing that an alternative feature was more urgently needed to address a critical pain point. By combining our expertise: yours in strategy and mine in user insights: we reinforce the credibility of our choices and ensure stakeholders feel confident in the process.

By working together to ground our decisions in research, craft unified narratives, and co-present findings, we can defend product decisions effectively while building trust with stakeholders. This not only strengthens our own collaboration but also ensures that our product meets both user needs and business goals.

Let's team together better by doing the following:

1. Collaborate on Research Plans

- Work together early to identify key questions and user insights needed for upcoming decisions.
- Align research objectives with business goals to ensure findings resonate with stakeholders.
- Partner on research activities like user interviews or usability tests to build a shared understanding of user needs.

2. Build Unified Presentations

- Partner to create a cohesive narrative for stakeholder presentations.
- Use visuals like journey maps or charts to clearly connect user insights to product decisions.
- Rehearse presentations together to ensure alignment on key messages and address potential questions.

3. Tie Decisions to Metrics

- Collaborate to link user feedback and research findings to business metrics like retention, conversion rates, or engagement.
- Highlight how decisions support broader company objectives to reinforce their strategic value.

4. Provide a Platform for Data

- Ensure stakeholder presentations include time to showcase user research and its connection to product decisions.
- Advocate for the importance of data-driven decisions during roadmap discussions or meetings with leadership.

#7
Cutting Through Subjectivity

How UX shift conversations
from opinions to evidence.

It must be challenging as a PM to navigate the sea of subjective opinions that often surrounds product decisions. Sales, marketing, and leadership each bring their perspectives on what should be built, and these perspectives can sometimes conflict with user-centered insights. My role is to help redirect these conversations toward data-driven decisions that deliver meaningful value to users while aligning with our business goals. By anchoring discussions in evidence, we can create clarity, reduce debates, and ensure our decisions have a measurable impact.

Using Data to Overcome Opinions

Grounding decisions in user research and usability testing allows us to move beyond subjective opinions. By focusing on real user needs, behaviors, and pain points, we can prioritize features and approaches with the greatest potential for impact.

For example, imagine a stakeholder proposes a feature aimed at boosting engagement, like adding gamification to a financial planning app. However, research and user interviews reveal that users are primarily focused on accuracy and simplicity when managing their finances. Users tell us they

are distracted, frustrated, and not motivated at all by prompts to collect stars or in-app points. By sharing these insights, and quotes from real users, we can redirect efforts toward enhancing other, more impactful features like error detection or intuitive navigation: improvements that better address user priorities and support retention and satisfaction.

Data transforms subjective debates into constructive discussions. When stakeholders see evidence of what users truly need, it becomes easier to align everyone around a shared vision. This not only improves the product but also reduces friction in the decision-making process.

Navigating Conflicting Feedback

Even with strong data, stakeholders may hold opinions that conflict with user insights. These moments require empathy and diplomacy. Acknowledging their perspectives is critical to building trust and fostering collaboration.

For instance, if a sales leader insists on prioritizing a flashy feature to attract enterprise clients, but research shows it's not a priority for existing

users, the response shouldn't be dismissive. Instead, we might say, "*I see how this feature could help with enterprise sales. Let's look at what our data says about user expectations and see how we can align the feature with their needs while achieving your goals.*"

By framing data as a shared resource rather than a rebuttal, we create space for dialogue and collaboration: and the opportunity to collect more data / conduct more user research. This approach helps balance user-centered insights with business needs, ensuring the final decision reflects both perspectives.

Experimentation as a Way Forward

When conflicting opinions arise, experiments like A/B testing or prototyping offer an objective path forward. These methods allow us to test both stakeholder ideas and user-driven alternatives, generating data that informs the final decision: a decision we can all be confident in.

For example, imagine a stakeholder proposes a complex dashboard layout for a new analytics tool, prioritizing maximum information density. At the

same time, usability testing reveals that users prefer simplicity and quick access to key metrics. Instead of debating which approach is better, we should offer to test both layouts with users, maybe including a third, hybrid solution. We then determine the approach that results in higher task completion rates and satisfaction, to create concrete evidence in support of the direction we choose. This process not only resolves disagreements but also fosters trust in data-driven methods.

Building a Culture of Evidence

Consistently incorporating research into conversations helps shift the culture toward evidence-based decision-making. Over time, stakeholders come to rely on user data as a trusted foundation for product decisions.

This cultural shift reduces reliance on gut instincts or personal biases and encourages a collaborative approach rooted in shared objectives. Stakeholders begin to build confidence in Product Leaders and see UX Research as a strategic asset, one that not only resolves debates but also drives better product and business outcomes.

By leveraging data to guide decisions, navigating conflicting opinions with empathy, and using experiments to validate ideas, we can ensure that product decisions are grounded in evidence rather than subjectivity. Over time, this approach builds trust, aligns teams, and creates products that deliver meaningful value to users and the business alike.

Here are things we can do to shift conversations from opinions to evidence, create a more collaborative, user-centered approach to decision-making that delivers better outcomes for both users and the business:

1. Encourage User-Centered Discussions

- Frame stakeholder conversations around user needs and data rather than opinions.
- Ask UX Designers to share insights that align with business goals during discussions.
- Reinforce the importance of addressing validated user pain points when prioritizing features.

2. Leverage Experiments to Resolve Disagreements

- Suggest A/B testing or prototypes to validate competing ideas with user data.
- Collaborate with UX to design experiments that provide clear, actionable outcomes tied to KPIs.
- Use the results to guide stakeholder discussions and align on evidence-based decisions.

3. Acknowledge and Align Stakeholder Perspectives

- Start by validating stakeholder concerns and demonstrating an understanding of their priorities.
- Work with UX to connect stakeholder goals to user research, highlighting shared objectives.
- Use data to suggest compromises that address both user needs and stakeholder vision.

4. Advocate for Evidence-Based Decision-Making

- Collaborate with UX to present research findings during roadmap discussions or stakeholder reviews.
- Emphasize the long-term value of prioritizing real, validated user needs over subjective preferences.
- Foster a culture where decisions are routinely grounded in data, reducing debates over personal opinions.

Using "Confidence Levels"

*A communication tool for discussing risk
and managing expectations.*

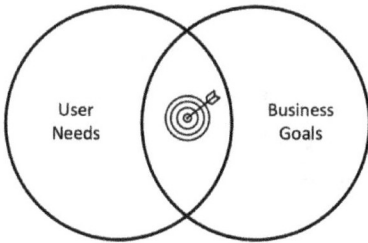

As UX Designers, we bring a rich blend of skills: interaction design, user research, usability testing, and visual design: to create products that are both user-friendly and business-aligned. When we have enough time to fully apply these skills, we can confidently deliver solutions that are well-researched and validated, minimizing risk and maximizing impact.

But let's be honest: tight deadlines and shifting priorities often mean we can't always rely on thorough research and testing. Sometimes, we have to lean on our expertise and intuition. This is where openly communicating our confidence in a design becomes essential. It helps everyone align, enabling smarter decisions about where to invest time and energy.

What Is a Confidence Level?

A Confidence Level is a simple, yet powerful way to convey how confident we are that a design will effectively meet both user _and_ business goals. It acts as a guide for making informed decisions and managing risk.

- **High Confidence:** The design is backed by solid research, testing, and data. It's been validated with users and directly supports business goals.

- **Medium Confidence:** Some research has been done, and the design likely meets user needs, but more testing could refine it. This is often a mix of lightweight validation and a designer's experience.

- **Low Confidence:** The design is based mainly on intuition, assumptions, or quick fixes, with little to no user testing. This carries a higher risk of missing the mark with users or meeting business objectives.

Balancing Risk with Confidence Levels

Obviously, each Confidence Level comes with various risks and trade-offs:

- **High Confidence = Low Risk, More Time**: Designs with high confidence are less risky because they're validated, but they require more time upfront for research and testing.

- **Medium Confidence = Balanced Risk**: A safer middle ground: good enough to move forward but can still benefit from future improvements.

- **Low Confidence = High Risk, Fast Delivery**: Speeds up delivery but increases the chance of needing costly rework later.

Focusing Effort Where It Matters Most

Not every feature needs exhaustive user research or the most polished design. Some features: especially those that follow standard, well known user interaction patterns: don't require extensive validation because users are already familiar with how they work.

Our goal shouldn't be perfection in every corner of the product. Instead, we should focus our energy on what matters most. High Confidence Levels should be sought after for critical product features, innovative designs that introduce risk, or user flows that directly impact key business metrics (KPIs). This approach ensures we're directing our time and effort into areas where design has the biggest impact, without getting bogged down in over-polishing low-risk features.

Using Confidence Levels to Guide Conversations

Sharing a Confidence Level isn't about labeling designs as "good" or "bad": it's about sparking meaningful conversations and communicating something most Designers and PMs don't discuss: **risk**. As a PM, your job is balance not just risk, but multiple risks from multiple perspectives: technical risk, schedule risk, and more. A Confidence Level is simply a way to communicate risk associated with design & usability, so you can balance that with the other pressures you face.

For example, if a design has a "Medium" Confidence Level due to limited testing, we can discuss whether it's worth investing more time in research or if it's acceptable to move forward as-is. Other risks may outweigh this design risk, but this transparency gives PMs and stakeholders a clear view of potential risks and allows the team to decide how to best proceed.

By framing our discussions around confidence, we can work together to make strategic decisions: Should we launch quickly and plan to iterate, or should we invest more time upfront to mitigate risk? These conversations help balance speed, quality, risk, and user needs.

Balancing Intuition with Validation

As experienced designers, we often have strong instincts about what will work for users. A low or medium Confidence Level doesn't mean a design is poor: it simply acknowledges that we've relied heavily on those instincts and our hypothesis haven' been fully validated yet. This distinction helps stakeholders understand where additional testing could strengthen the solution.

For example, a quick redesign of a dashboard might streamline navigation, but without user testing, we might miss critical pain points like confusing error messages. Pairing our intuition with user testing in future phases could allow us to improve the Confidence Level up to "Medium" in order to balance delivery speed with thoughtful design improvements.

Handling Low-Confidence Situations with Transparency

When tight deadlines force us to move forward with low-confidence designs, honesty and collaboration become even more important:

- <u>Clearly Communicate Risks</u>: We should be upfront with stakeholders about assumptions and risks. For example, *"Without user testing, we're unsure how users will interpret this feature, which could lead to confusion."*

- <u>Propose Phased Approaches</u>: We should consider launching a basic version to meet deadlines, with a plan to gather feedback and iterate. For instance, releasing a simplified dashboard first, then adding advanced features based on user input.

- <u>Use Confidence Levels to Drive Discussion</u>: Confidence Levels give us a common language to talk about risks and trade-offs. Saying, *"This is a medium-confidence design because it hasn't been user-tested yet"* invites the team to consider if more testing is worth the time.

This transparency fosters trust and helps everyone feel more comfortable balancing tight timelines and shifting priorities.

Again, not every design needs perfection, but critical features, risky innovations, and KPI-driving flows deserve deeper validation to reduce risk and drive success.

Confidence Levels help us focus our time and energy where it matters most and here our effort are most impactful to the product's success.

By using Confidence Levels, we can align expectations, manage risks, and make smarter decisions as a team. This approach allows us to balance speed with quality: ensuring that what we deliver meets user needs and business goals without wasting time perfecting things that don't need it.

Here are some steps we can take to start weaving Confidence Levels into the way we work:

1. **PMs: Ask for Confidence Levels During Design Reviews**

    - Proactively invite UX Designers to share a Confidence Level when presenting their work.
    - Ask questions like, *"How confident are we that this design will meet user needs and business goals?"*
    - This opens the door for honest discussions about risks, validation, and next steps.

2. **Prioritize Time for Research on High-Impact Features**

    - Collaborate with UX to identify critical features or user flows that directly impact KPIs.
    - Allocate extra time and resources for user research and testing on these high-priority areas to increase design confidence where it matters most.

3. Use Confidence Levels to Guide Roadmap Decisions

- Incorporate Confidence Levels into roadmap and sprint planning discussions.
- Use them to weigh the trade-offs between speed and quality, helping the team decide when to move forward and when to invest in further validation.
- Leverage Confidence Levels to aid decision-making in the same way you would use other prioritization frameworks (e.g., MoSCoW, Kano Model, or RICE scoring).

4. Encourage Iteration Through Phased Releases

- When launching lower-confidence designs create a clear plan for follow-up improvements.
- Advocate for phased releases that allow for user feedback and iterative design updates, reducing risk while maintaining momentum.

Contributing Value:
What and When

How UX delivers value across product phases.

As your UX Designer, I can provide a wide range of artifacts, services, and deliverables tailored to each phase of feature development or product development. These outputs aren't just "nice-to-haves"; they are tools that can directly contribute to the product's success by informing decisions, validating concepts, and optimizing the user experience.

Below, I've outlined *what* I can create during each phase of development, along with *how* you can use these deliverables to align with business goals and user needs. Feel free to ask for any of these based on where we are in the process!

Discovery/Research Phase

During this phase, we lay the foundation for the product by understanding the problem space, user needs, and market opportunities.

- **User Research Reports**

    - **Use:** To gain insights into user behaviors, pain points, and motivations, helping shape product strategy and feature prioritization.

    - **Example:** A report revealing that users struggle with onboarding due to unclear instructions can inform prioritizing a simpler onboarding process.

- **Personas**

  - **Use:** To create detailed representations of target users, fostering empathy and ensuring design decisions are user-focused.

  - **Example:** A persona for "Busy Professionals" highlights the need for time-saving features, guiding design and feature prioritization.

- **User Journey Maps**

  - **Use:** To visualize the end-to-end user experience, identifying touchpoints and pain points.

  - **Example:** Mapping the journey of a first-time app user reveals frustration during account setup, leading to changes that reduce friction.

- **Competitive Analysis Reports**

    - **Use:** To identify strengths, weaknesses, and opportunities by evaluating competitor products.

    - **Example:** Learning that a competitor's search feature outperforms ours might prioritize improving our search functionality.

- **Information Architecture (IA) Diagrams**

    - **Use:** To structure content logically, making navigation intuitive.

    - **Example:** An IA diagram for an e-commerce site ensures that categories are easy to find and relevant to users.

Design Phase

This is where the vision takes shape through conceptualization, iteration, and validation.

- **Wireframes**

  - ○ **Use:** To outline the structure and layout of screens without getting bogged down by detailed visuals.

  - ○ **Example:** Wireframes of a dashboard help align stakeholders on basic functionality before moving to detailed design.

- **User Flow Diagrams**

  - ○ **Use:** To map the steps users take to accomplish tasks, identifying potential pain points.

  - ○ **Example:** A flow diagram shows a convoluted checkout process, prompting a redesign to streamline purchases.

- **Prototypes**

  - **Use:** To create interactive models of the product for usability testing and stakeholder feedback.

  - **Example:** A clickable prototype of a mobile app lets users test navigation, revealing confusion around menu placement.

- **Usability Testing Reports**

  - **Use:** To gather feedback on prototypes, highlighting issues to address before development.

  - **Example:** Testing a search function uncovers that users struggle with filters, prompting improvements.

- **Style Guides and Design Systems**

  - ○ **Use:** To ensure consistent visuals and interactions across the product lifecycle.

  - ○ **Example:** A design system includes reusable components like buttons, reducing developer guesswork and speeding up development.

Development Phase

As development ramps up, UX ensures the design vision translates into reality.

- **Final Design Assets**

    - ○ **Use:** High-fidelity mockups and design files provided to engineering teams ensure accuracy during development.

    - ○ **Example:** Pixel-perfect assets for a marketing page ensure the live version matches the design.

- **Interaction Design Specifications**

    - ○ **Use:** To detail how animations and transitions behave, clarifying expectations for developers.

    - ○ **Example:** Specifying how a menu expands or collapses prevents inconsistent implementations.

Pre- and Post-Launch Phases

Once the product is built, UX can validate its success and inform future iterations.

- **Heatmaps and Analytics Reports**

    - **Use:** To analyze user engagement and identify areas for optimization.

    - **Example:** A heatmap reveals users rarely scroll below the fold, prompting a redesign to surface key content earlier.

- **User Feedback Surveys and Reports**

    - **Use:** To gather direct user input, identifying satisfaction levels and areas for improvement.

    - **Example:** A survey reveals that users want more customization options, informing future updates.

- **A/B Testing Results**

    - **Use:** To compare different versions of features, determining which performs better.

    - **Example:** Testing two button designs shows that one increases clicks by 15%.

Throughout the Product Lifecycle

UX support doesn't stop after launch. We continue to add value by iterating based on user feedback and evolving goals.

- **Roadmaps and Feature Prioritization Documents**

  - **Use:** To plan future updates and ensure alignment with user needs and business objectives.

  - **Example:** A roadmap that prioritizes features addressing high-impact user pain points keeps the team focused on value delivery.

#10
You Need More Than a Designer

How a Design Methodology saves time and drives results.

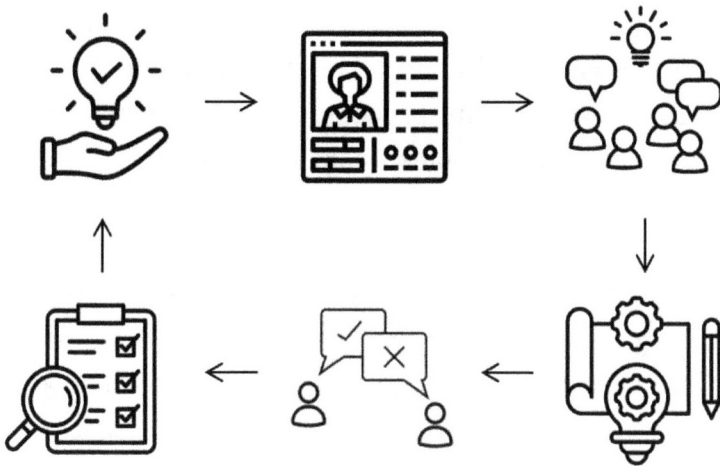

When a product struggles with user adoption, isn't very usable, or isn't meeting its success metrics, many companies turn to hiring a UX Designer to "fix it." And while a talented designer can certainly make improvements, relying on one person, and only one person's perspective often leads to incomplete solutions that only scratch the surface of the problem.

The best approach is to adopt a Design Methodology: a better way of working, and a structured way of thinking and solving problems. Two great examples are: **Design Thinking** and the **Double Diamond** framework. These approaches ensure teams of people with different perspectives work collaboratively, explore multiple solutions, and validate their ideas, ultimately creating more impactful and successful products.

This chapter offers an introduction to these methodologies, but it's just the tip of the iceberg. Design is a vast discipline, and these frameworks are starting points. If this chapter piques your interest, I encourage you to dive deeper into other books, websites, and resources to fully understand and apply design methodologies.

Why a Designer Isn't Enough

It is a common misconception that a single designer, or a any one person at all, can single handedly solve a product's usability issues. While UX Designers can bring valuable skills to the table, placing the responsibility solely on one person is severely limiting. Every UX Designer's ideas and solutions are shaped (and limited) by their individual experiences, previous projects, jobs, roles, as well as the constraints they face. A Designer's first idea usually isn't great! It may not be the best or most impactful solution for the product and its users. A more structured process of exploring and testing multiple ideas is needed to increase confidence.

For example, imagine a struggling product with low user retention. A designer might suggest improving the product's visual appeal or simplifying a specific interaction. However, without a methodology that leads with empathy to understand users deeper issues: such as unclear onboarding or misaligned product positioning: the core issues might remain unaddressed. A design methodology prompts not only a deeper understanding of the real problem, but then cross-functional collaboration to explore multiple options, multiple directions, validate assumptions, and

prioritize solutions that address root causes rather than surface-level problems.

Understanding Design Thinking and Double Diamond

Two widely adopted design methodologies, **Design Thinking** and **Double Diamond**, offer structured approaches to problem-solving that go far beyond a single designer's input and intuition.

Design Thinking

This human-centered methodology involves five stages:

1. **Empathize**: Understand users' needs through research, interviews, and observation.
2. **Define**: Clearly articulate the problem based on research insights.
3. **Ideate**: Brainstorm multiple solutions with divergent thinking.
4. **Prototype**: Create low-fidelity representations of solutions to test ideas quickly.
5. **Test**: Gather user feedback to refine and validate solutions before implementation.

For example, consider a team tasked with improving an e-commerce checkout process. During the Empathize stage, they discover users abandon carts due to confusion about shipping costs. In the Define stage, they focus on this specific problem. Ideation leads to multiple ideas like cost previews, bundling incentives, and real-time updates. Prototyping and testing these ideas reveal that a "cost calculator" is the most effective solution, which is then refined and implemented.

Double Diamond Framework

This model divides the design process into four phases:

1. **Discover**: Explore the problem space through research and user feedback.
2. **Define**: Narrow down to the core problem based on insights.
3. **Develop**: Create and test multiple prototypes or concepts.
4. **Deliver**: Finalize and launch the best solution.

For instance, a team improving a fitness app might discover during the Discover phase that users struggle with tracking progress. The Define phase narrows down the problem to their confusion about how long it will take to reach the goals they set. In the Develop phase, they prototype a few ideas including visual dashboards, gamified goals, and customizable summaries. User testing in the Deliver phase confirms that a better dashboard meets user needs and drives higher engagement.

Benefits of a Structured Design Methodology

1. **Encourages Exploration**: These methodologies prioritize *divergent thinking*: exploring multiple solutions before converging on the best option. This reduces the risk of investing in ideas that don't work.
2. **Aligns Teams**: Cross-functional collaboration ensures that user needs, technical constraints, and business goals are integrated into the solution.
3. **Saves Time and Resources**: Testing ideas early prevents costly rework later. For instance, building a prototype cost far less than coding an unvalidated feature.
4. **Drives Innovation**: By engaging diverse perspectives and testing multiple solutions, teams are more likely to uncover innovative ideas that resonate with users.

Here are some steps we can take to embrace Design Methodologies:

1. Prioritize Problem Definition

- **When:** At the beginning of any project.
- **How:** Work to frame the problem clearly before jumping to solutions. Ask, *"What are we solving for?"*
- **Why:** Defining the problem ensures you address the right challenges, reducing wasted effort on misaligned solutions.

2. Facilitate User Research

- **When:** Early in the project and throughout the process.
- **How:** Conducting interviews, surveys, or usability tests.
- **Why:** Grounding the process in real user insights ensures solutions address actual needs, not assumptions.

3. Advocate for Iteration

- **When:** During planning and design phases.
- **How:** Build time into the schedule for prototyping, testing, and refining solutions.
- **Why:** Iteration leads to higher-quality outcomes and reduces risks before development.

4. Promote Cross-Functional Collaboration
- **When:** Throughout the product lifecycle.
- **How:** Organize workshops or brainstorming sessions that include designers, developers, and stakeholders.
- **Why:** Engaging diverse perspectives ensures solutions are holistic and feasible.

5. Explore and Invest in Design Training

- **When:** Ongoing.
- **How:** Provide opportunities for the team to learn more about design methodologies, whether through workshops, books, or online courses.
- **Why:** A team that understands design processes can collaborate more effectively and value the role of UX in product success.

By adopting a Design Methodology, we're not just enhancing the role of your UX Designer: we're embedding a process that ensures our product is user-centered, strategic, and innovative.

NOTE: While this chapter introduces foundational concepts, there's much more to explore! Seek out additional resources to deepen your understanding of various Design Methodologies to get the most out of UX for your Product... creating products that don't just look great, but deliver real, lasting value.

Conclusion

The partnership between a Product Manager and a UX Designer is about more than just collaborating on a product: it's about leveraging the unique strengths of both roles to solve problems effectively and create exceptional results.

Throughout this book, we've explored many ways PMs and UX Designers can partner together far beyond look, feel, and aesthetics. From helping shape product roadmaps to cutting through subjectivity and driving data-driven decisions, UX: as a problem-solving discipline: can offer insights, communicate risk, help prioritize efforts, reduce cost, and align user needs with business goals. By engaging UX Designers early and integrating this expertise into every stage of development, you empower them to help elevate the product and drive its success.

This book is one half of a larger conversation. For a PM and UX Designer to truly thrive together, both roles must embrace empathy, curiosity, and collaboration. This isn't about pointing fingers or delegating blame: it's about understanding and supporting each other to build great products together. When trust and open communication are the foundation of your partnership, the results will speak for themselves.

As a Product Manager, embracing the iterative nature of design, integrating your UX Designer early, and making space for user insights will enhance the collaborative process. Whether it's leveraging Confidence Levels, validating decisions with user research, or iterating on solutions before

launch, these tools strengthen decision-making, build trust with stakeholders, and ensure balance across user and business goals.

Ultimately, the goal is to be able to leverage UX Designers for more than just visual design, tapping into their ability to be a strategic partner in delivering impact. By applying insights from this book, you'll build a stronger partnership that leads to better products, greater success for your team, and more meaningful outcomes for your users.

Together, through shared problem-solving and collaboration, you can create products that not only meet deadlines but also stand out in the market for their thoughtful, user-centered design.

Bonus: A PM-UX Partnership Manifesto

10 Core Principles for Collaboration

What is a PM-UX Partnership Manifesto?

A manifesto is a shared set of principles aimed at strengthening collaboration between Product Managers (PMs) and UX Designers (UX). It outlines the values, practices, and mindsets that help both roles thrive together, ensuring that business goals and user needs are met harmoniously. Whether you're starting a new project or refining your existing processes, these principles serve as a practical guide to build trust, foster collaboration, and create better products. Use them in team discussions, roadmap planning, design critiques, or retrospectives to align your efforts and maximize your impact.

By embracing these principles, PMs and UX Designers can form a partnership that not only delivers exceptional products but also fosters a collaborative, innovative team culture. Together, we can create solutions that meet user needs, drive business success, and set a new higher standard for teamwork.

1. Shared Ownership of Product Success

We are united by a common goal: creating products that deliver value to users and drive business success. UX Designers and PMs share responsibility for achieving this outcome, with each bringing unique perspectives and skills to the table. Together, we solve problems, not just delegate tasks.

How to Apply: Start every project with a shared definition of success. Align on key metrics, user goals, and business objectives. Revisit this definition regularly to ensure focus.

2. Empathy for Users and Each Other

Empathy is at the heart of what we do: not just for our users, but for each other. PMs respect the time, expertise, and insights that UX Designers need to craft thoughtful solutions. UX Designers understand the constraints and pressures PMs face in balancing timelines, stakeholders, and goals.

How to Apply: Schedule regular check-ins to share challenges, insights, and updates. Use these moments to build mutual understanding and address roadblocks together.

3. Start with Problems, Not Solutions

We collaborate best when we frame our work around clear problems rather than predetermined solutions. By defining the "why" before the "what," we ensure that our efforts address meaningful user needs and business priorities.

How to Apply: Use problem-definition workshops, brainstorming sessions, or user journey mapping exercises to align on challenges before jumping into solutions.

4. Data-Driven Decision Making

We prioritize decisions grounded in data: user research, usability testing, and analytics: over assumptions or opinions. When data isn't available, we agree to test hypotheses and validate decisions through iterative processes.

How to Apply: Build time for research, prototyping, and testing into your roadmap. Use tools like Confidence Levels or A/B tests to evaluate and refine ideas collaboratively.

5. Embrace Iteration and Exploration

The best solutions often emerge through exploration and iteration. We value the process of testing, learning, and refining over rushing to deliver the first idea. Multiple perspectives and feedback loops lead to better outcomes.

How to Apply: Allocate time for designers to explore multiple concepts and for PMs to review and discuss them. Plan iterative development cycles that include user feedback.

6. Transparent Communication

Clear, open, and proactive communication is the foundation of our partnership. We share context, goals, constraints, and progress openly to build trust and avoid surprises.

How to Apply: Use shared tools like project briefs, research repositories, or collaborative roadmaps to maintain visibility. Host regular syncs to align on priorities and expectations.

7. Respect Each Other's Expertise

PMs trust UX Designers to advocate for users and deliver high-quality experiences. UX Designers trust PMs to align the product vision with business goals and stakeholder needs. Each role values the other's expertise and contributions.

How to Apply: Involve each other early and often in discussions about priorities, strategy, and design. Acknowledge contributions and give credit for successes.

8. Focus on the End-to-End Experience

Great products consider the entire user journey, not just isolated features, or screens. We collaborate to design workflows and solutions that are seamless, intuitive, and aligned with the broader product vision.

How to Apply: Approach projects through the lens of user flows and systems thinking. Use tools like journey maps or experience maps to identify opportunities and gaps.

9. Advocate for Users, Advocate for Business Goals

Our partnership thrives when we strike the right balance between user needs and business objectives. UX Designers advocate for the user experience, while PMs ensure alignment with broader company goals. Together, we champion both sides.

How to Apply: Frame discussions around both user impact and business outcomes. Use user stories, metrics, and real-world examples to connect the two.

10. Foster a Culture of Collaboration

Collaboration isn't just a process: it's a mindset. We approach our work with curiosity, humility, and a commitment to learning from each other. We embrace feedback and view every interaction as an opportunity to strengthen our partnership.

How to Apply: Host team retrospectives to reflect on what's working and what's not. Celebrate wins together, no matter how small, and recognize the value of cross-functional contributions.

Practical Applications for Teams

- **In Kickoff Meetings**: Review the manifesto together to align on expectations and collaboration practices.

- **During Retrospectives**: Use the principles as a framework for evaluating team dynamics and identifying areas for improvement.

- **For Onboarding New Team Members**: Share the manifesto to introduce your team's approach to collaboration and the value of the PM-UX partnership.

- **As a Reference Tool**: Keep the manifesto visible: whether in a shared workspace or pinned in your collaboration tool: so it's always top of mind.